MEMORIES OF THE East Anglian Fishing Industry

MEMORIES OF THE East Anglian Fishing Industry

Ian Robb

COUNTRYSIDE BOOKS
NEWBURY BERKSHIRE

First published 2010

COUNTRYSIDE BOOKS
3 Catherine Road
Newbury, Berkshire

To view our complete range of books,
please visit us at
www.countrysidebooks.co.uk

ISBN 978 1 84674 208 8

Designed by Peter Davies, Nautilus Design
Produced through MRM Associates Ltd., Reading
Typeset by Mac Style, Beverley, UK
Printed by Cambridge University Press

Contents

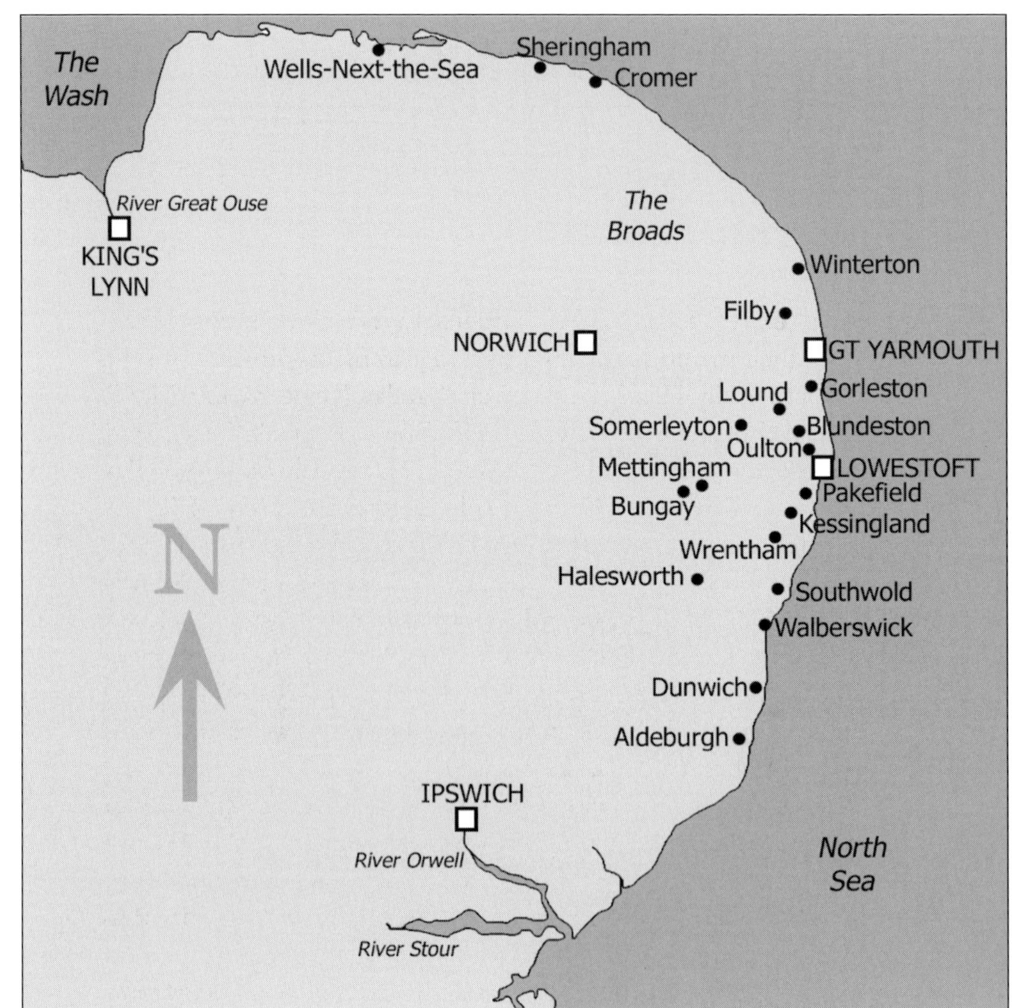
The Wash
Wells-Next-the-Sea
Sheringham
Cromer
River Great Ouse
KING'S LYNN
The Broads
Winterton
Filby
NORWICH
GT YARMOUTH
Lound
Gorleston
Somerleyton
Blundeston
Oulton
LOWESTOFT
Mettingham
Bungay
Pakefield
Kessingland
Wrentham
Halesworth
Southwold
Walberswick
N
Dunwich
Aldeburgh
IPSWICH
River Orwell
River Stour
North Sea

Acknowledgements

I cannot express enough my sincerest gratitude to everyone who shared their memories of the East Anglian fishing industry for this book, especially, Hazel Arnold, Vera Barclay, Kevin Barrett, Mrs E. Blaney, Hazel Bliss, Brian Blowers, Peter Boyce, Reginald Brine, Mark Butcher, Ernie Childs of the Great Yarmouth Pottery, Glen Chipperfield, Rosemary Crisp, Ernie Dewhurst, Doug Ellis of Ipswich, Tom Field, Rita Flatt, village historian at Lound, Suffolk, John and Maureen Fryers, Paul Hoey, Ken and Iris James, Alan Jones, Mike and Brenda Leggett, Mrs Betty Meades, Mike Moyes, Jim and Jeanette Nolloth, Frank and Elinore Philpot, Stuart Philpot, Chris Pink, Beryl and George Read, Alan Stevens, Albert Stone, and Ron Wilson. My apologies to those whose memories I was unable to use in this book, this was solely because of lack of space.

Thanks also to the following who allowed me access to their written and photographic material: the Jack Rose collection in the Ian G Robb Photographic Archive, Pam Graystone, Ernie Childs, Paul Allison, Peter Killby, Frank Philpot, Hazel Bliss, Beryl Joan Read, Chris Pink and Paul Hoey.

Many thanks to the following for their help and encouragement: Suffolk Record Office at Lowestoft; Ernie Childs, Great Yarmouth Pottery; Mark Butcher; Jonathan Winterton; Suffolk County Councillor Sue Allen (Southwold and Reydon local historian); Paul Allison; Bob Malster who told me about True's Yard Museum at King's Lynn; members of the Jack Rose Old Lowestoft Society; *East Anglian*

Daily Times; *Yarmouth Mercury*; *Eastern Daily Press*; *Lowestoft Journal*, and Leslie Dolphin at BBC Radio Suffolk.

Finally, sincerest thanks to the following museums, to which I strongly recommend a visit by anyone who would like to know more about the East Anglian fishing industry: Lowestoft and East Suffolk Maritime Museum; Time and Tide Museum, Great Yarmouth; The Old Smokehouse at Great Yarmouth Pottery; True's Yard Museum, King's Lynn.

Introduction

'The fishermen used to say, "Whatever else, we've always got the fishing".'

Maureen Fryers, Lowestoft

When I, Ian Robb, entered the East Anglian fishing industry in 1962, it was as an eager 15½-year-old tally boy at Explorator. Explorator was a wholesale fish merchant which had been founded in Lowestoft in 1947. As a tally boy, or to be more correct, a tally clerk, I was responsible for writing out the customer destination and contents labels – known as tallies – which were then affixed to the boxes in the packing station below; the offices being on the first floor. There were generally three of us young boys in a small office at the back of the building. Over the years we must have written out hundreds of thousands of tallies, if not millions!

I came from a long-established fishing family, the Capps, and believed I was continuing in a tradition that would undoubtedly carry on well after I had gone. Like many of my generation, I went straight from school into fishing – almost guaranteed a job for life. Granted, there had been the occasional depression and periodical drop in demand down the years, the last being in the 1930s, but the East Anglian fishing industry looked set to prosper well into the next century. Following the 1964 herring season, I asked in my youthful zeal: 'with such an abundance of herring being caught this year, what if next year they disappeared?' I was considered somewhat eccentric for this thought. Regretfully, however, 1965 did indeed see the demise of the herring, once hailed as the 'Silver Darling'.

The fishing industry at one time involved communities great and small on the Norfolk and Suffolk coast, ranging from King's Lynn and the River Ouse mainly concentrating on the Wash in the north-west, to Ipswich on the Orwell in the south. It also included two of the largest ports on the east coast – Great Yarmouth and Lowestoft – as well as a number of smaller ports once renowned for specialities as diverse as plaice, cod, sprats, and shellfish. This East Anglian affinity with the sea also gave birth to many heroes of the Royal Navy: in the 17th century Sir Thomas Allin and Sir Christopher Myngs, and of course, the greatest of them all, in the 18th century, Vice-Admiral Horatio Nelson.

We are fortunate that fishing has left us with an important legacy and we need to preserve the memories and stories of those who worked in this industry. Southwold's part in the annual Home Fishing, which took place before the First World War when Scots boats in the shape of Zulus and Fifies bore down en masse on the little harbour laden full of herring, has long gone, only to be kept alive by photographs. King's Lynn's own North End, an area minutes away from the Tuesday Market Place, housed the port's fishing community until it succumbed to the clearances and redevelopment of the 1930s and 50s. Those who recall the annual migration of the Scots fishergirls to gut, pickle and pack the herring in Lowestoft, Great Yarmouth and Gorleston will undoubtedly disappear, to be limited to the few who were children in the 1950s and 60s. And the local delicacy of cockles, whelks, lobsters and shrimps once caught in the Wash and off the north Norfolk coast, which were noted for their particular 'seaside' taste, reminiscent of balmy days on the beach and walks along the pier at Great Yarmouth, Lowestoft, Southwold or Cromer, will be supplanted by quick-frozen imports.

Many of the East Anglian fishermen could not believe that the herring would disappear in their lifetime – but that is exactly what happened. However, nothing could prepare them for what would happen in the final decades of the 20th century – the end of a way of life and the almost complete annihilation of an entire industry well over a thousand years old, caused firstly by over-fishing and later by governmental red tape and the high cost of fuel.

For much of their existence, the ports of Norfolk and Suffolk benefited from their location close to or on major sea routes. They connected the north-eastern ports of England with London, and also attracted foreign traders. In addition, apart from King's Lynn and Ipswich, the ports were located close to the North Sea fishing grounds and the migrational route of the herring – a fish which never changed its route or habits. It was the Silver Darlings – the herring - that caused the East Anglian fishing industry to grow from boats fishing for a local community, to supplying coastal towns that were, by the Norman invasion of 1066, paying their dues and demands mainly in fish. Fish were used to supplement the lack of fresh meat as part of a staple diet during the winter months. The herring served as part of a religious diet, especially during Lent, and continued to do so until the Dissolution of the Monasteries began in 1536.

For the most part East Anglian fishermen used drift nets – ideal in such shallow waters as the North Sea – and which, despite periodical over-fishing, allowed fish stocks to continue to flourish for the following season. Extensive North Sea trawling, on the other hand, arrived late in East Anglia, around 1850, and followed the arrival of the railway in the mid-19th century when it became possible to deliver fish to customers as far afield as London and the Midlands within twenty-four hours of landing it. A hundred years later and stocks of most species in the North Sea were becoming exhausted as heavy fishing by bottom trawling disturbed the spawning fish.

Looking back today at what was once a great fishing industry, an industry which had survived over a millennium, its sailors battling against sea and storm to bring home the harvest of the sea, it is astonishing just how quickly it all disappeared. Within the last two decades of the 20th century, fishing in the North Sea as I once knew it, and which my parents and grandparents took for granted, disappeared, never to return.

Today, all that remains of this once great fishing heritage are the vestiges of an industry which, in its time, made this part of England not only a lively and prosperous region, but also one that gave much of the East Anglian coast its unique character. The culture that once developed around East Anglia's fishing communities has given way to new industries. Some communities managed to adapt and begin anew in other fields of industry – oil and gas, for instance, and more recently offshore wind farms; sadly, others never fully recovered.

Chapter 1

A Short Tour of the Ports

We begin in north Norfolk, at the sea port and market town of King's Lynn. The town stands on the River Ouse and is approximately ten miles from the Wash, which leads into the North Sea. The port became an important mercantile centre after the Norman Conquest and traded with the German Hanseatic ports. It even remained busy during the Napoleonic Wars, when it continued trading with the Baltic, Spain and Portugal. In the late 18th century, King's Lynn was a large, rich and busy town. Indeed, the Census of 1801 suggests that Lynn was the second largest Norfolk port after Great Yarmouth.

Although fishing on the Ouse, the Wash and the North Sea played an important part in the town's economy, looking through early directories, such as the 1794 edition of the *Universal British Directory*, very little is said of the town's fishermen apart from the mention of a fish market on the Tuesday Market Place. In fact, the Tuesday Market Place was not far from Lynn's bustling fishing community, the North

End. A thousand years of Lynn's fishing families centred on this one small area, where practically everyone was involved in the fishing industry. From around 1185 to the early 20th century, when much of the area was demolished, the North End grew around the north and east sides of St Nicholas, the chapel of ease to St Margaret's parish church.

In the 1950s, King's Lynn fishermen landed their catches, which included cod, sole and shellfish, at the Fishers Fleet dyke which they had done since at least Elizabethan times. Herring were also caught and smoked by local fishermen. The Wash was a rich source of whelks, cockles and mussels, and until the 1970s, a significant proportion of British shellfish came from this part of the coast. Lynn's decline as a shellfish centre is said to be due to the deteriorating conditions of the seabed in the Wash and off the north Norfolk coast.

The production of salt from seawater has been an important industry, mentioned as far back as the Domesday Book of 1086. It is an essential part of fish production, and was probably – apart from air-drying – the earliest method of preserving fish, especially herring.

Cromer is situated further along the coast as you head towards Great Yarmouth. The shoreline here is rocky and can be treacherous. Over the centuries, ships have come to grief off the coast, which in the days of sail was once described as the *terror of sailors*, with Cromer bay the *Devil's Throat.* Despite this reputation, the town was inhabited mainly by fishermen, many of whom fished for lobsters which were sent overland to Norwich. The *Universal Directory* makes a note of the seasonal catches off Cromer – crabs and lobster between May and October and herring in September and October. Other fish found off its coast included turbot, codling, whiting, haddock and flounders. Crabs are still a vital industry today.

Until the arrival of the lifeboat, there was also a lucrative trade in salvage, for, like Lowestoft on the Suffolk coast, many buildings in and around Cromer were described as having – again quoting the *Universal Directory*; 'Houses, barns, stables and sties built of ships' planks, beams … from ships wrecked [off the coast].'

Both Cromer and the nearby resort of Sheringham have a reputation for shellfish, and predominantly crabs. In the 20th century, it became a tradition with local fishermen to sell directly to the public during the summer season, as Rosemary Crisp of Kessingland recalls:

> 'During the late 1940s and early 1950s our two weeks' summer holidays were spent at Aunt Elisabeth and Uncle Jack Gutteridge's house, *Dumbro House*, in Sheringham. Auntie was a good cook and insisted on using the freshest ingredients possible … Early in the morning Auntie would give me a large basket and my two brothers and I would walk down to the seashore, find Auntie's favourite boatman and buy some crabs. He would place the wriggling claw-snapping sea creatures in my basket. On the walk home the crabs would climb up the side and my brothers would push them down to the bottom.
>
> At Auntie's I would place the basket on the large well-scrubbed kitchen table and run out of the room. Auntie had a large pot of boiling water … on the shiny black-leaded stove. Us youngsters used to hate to hear the horrible sound that the crabs made when they were immersed in the pot. We ate the crabs at teatime; they were sweet, delicate and delicious!'

The Sheringham fishermen closely guarded the sites of their shellfish beds. So much so, it seems that, according to Suffolk historian Bob Malster, several Sheringham fishing families moved down to Felixstowe to try their luck there instead.

At the end of the 19th century, a small shellfish trade had also developed off the coast at Great Yarmouth. The Yarmouth shrimp was a particular favourite with the town's visitors during the holiday

Sheringham fishermen, c.1904 – note the lobster and crab cages.

season, which ran from May to August. They were caught using side trawl nets some four and a half metres long. A smaller version of the industry found elsewhere on the Norfolk coast, at Yarmouth the shrimps were usually boiled at home. Approximately 1,000 gallons per day were reputed to have been eaten by holidaymakers at the resort at the height of the season. If I remember correctly – I was only six or seven years old at the time – the industry still existed in the 1950s. I recall family trips to Yarmouth, especially to a particular public house on Marine Parade (I don't remember its name) where there was a booth in the beer-garden selling a number of shellfish delicacies – my favourite was a six-penny plate of cockles!

Located on a sandbank at the mouth of what was once known as the Great Estuary, Great Yarmouth describes itself, quite rightly, as the town that grew out of the sea. As early as the Domesday survey of 1086, Yarmouth had already established itself as a fishing and mercantile community. The sandbank's location, which eventually clogged up the Great Estuary, placed the founding port at the mouth of an inland sea that also served the important city of Norwich. This, and the obtaining of its charter in 1209, helped to make Yarmouth the wealthiest port in East Anglia. Both Yarmouth and later its neighbour, Lowestoft, were destined to play a significant part in the East Anglian fishing industry, and in Chapter 3 we will look at how this led to great rivalry between the two towns.

Whereas Yarmouth started life as a sandbank at the mouth of the Great Estuary – now the River Yare and the flatlands from which the Bure, the River Waveney and the Norfolk Broads were formed – the future fishing port of Lowestoft, in Suffolk, arrived late on the scene. It was founded on a cliff top protected from the sea by the Denes, a broad stretch of sand and marram grass which at one time may have stretched along the coast as far as Corton. Lowestoft was originally founded approximately half a mile inland, near to the ancient manor of Akethorpe and St Margaret's church. Originally an agrarian community, the terraces on the cliff side of the town appear to have been laid out simultaneously, suggesting it may have been established in the 13th century after Great Yarmouth received its charter from King John.

Lowestoft consisted of one main street interspersed with lanes on the landward side, and scores – mainly natural indentations in the cliff – on the seaward side. Shipbuilding and repair, curing, fish houses, storing, making, repairing and drying nets – in fact anything to do with the sea and with fish – were all done at the foot of the cliffs and on the Denes. With a considerable width between the sea and the cliffs, by the 18th century Lowestoft Denes also acted as a ropewalk, drying ground, and – if the weather allowed it – a cricket pitch during the summer months! Unlike Yarmouth, until 1830 Lowestoft had no harbour or haven. A bank of sand and shingle in front of the Denes allowed fishermen and merchants (occasionally one and the same) to beach their craft there. Cargo was usually rowed to land from boats moored offshore.

It was the arrival of the entrepreneur Sir Morton Peto, in the 1840s, that eventually saw Lowestoft's rise as a seaside holiday resort. Then living in Norwich, he was responsible for the branch line from Lowestoft to Reedham thereby linking the Lowestoft fishermen with their markets in the Midlands. It was also this railway that enabled Peto to develop what in time would become south Lowestoft. The existing harbour and docks were developed by the Eastern Counties Railway from 1848 and later, from 1862, by the Great Eastern Railway. This enabled the town to become an important trading port with the Continent and saw Lowestoft develop into a holiday resort, fishing port and market.

Travelling down the coast from Lowestoft, we arrive at Southwold. This is an ancient port that was once much larger, but has been reduced in size by coastal erosion, a problem faced by many communities on the Norfolk and Suffolk coast. Fishing at Southwold dates back to Domesday when the town was held as a manor for the monks of St

Edmund's; fishermen there annually caught 25,000 herring to feed the monastery. The town received its charter from Henry VII in 1488, following the deteriorating state of the port of Dunwich which had been established as a port of some importance in the 11th century. Dunwich went from being one of the largest ports in Eastern England during the 12th and 13th centuries, to a small coastal village, as eighteen churches and chapels, as well as most of the houses were lost to coastal erosion. Like most fishing ports along the East Coast, the Dissolution of the Monasteries between 1536 and 1541, and the subsequent abandonment of church fast days temporarily curtailed Southwold's fishing activities. A great fire in April 1659 destroyed much of the town, including the town hall and market place, as well as shops, granaries, warehouses and the homes of most of the town's fishermen, leaving many families facing poverty. Southwold's history more than once involved the Dutch herring fleet, and like Lowestoft, it was also involved in the Dutch Wars of the 17th century, which for Southwold culminated in the Battle of Sole Bay, fought off its coast in 1672.

By the 1790s, the town's trade was mainly in salt, brewing, herring and sprats. The herring fishery continued into the 19th century. Sole Bay was still considered a good anchorage in the late 18th century, sheltered as it was from the northern winds by the Easton Ness headland. The sprat industry made a temporary return during the interwar years of the 20th century.

In 1792, at the start of the French Revolution, Southwold was a pleasant, busy town. It was defended by guns on the cliff and almost surrounded by the River Blyth and entry was via a drawbridge. It was therefore considered to be in a strong position if attacked by the French. While the sea was washing away the coastline elsewhere, Southwold, like Great Yarmouth, had the reverse problem with sandbanks clogging its haven mouth. Despite this latter difficulty there were 204 'registered' fishermen listed in 1831, out of a population of 1,875.

Aldeburgh today is best remembered as the home of the Lowestoft-born composer, Benjamin Britten, for its art galleries and the annual Aldeburgh music festival; however, like Southwold, it too is an ancient seaport. Its rise was probably also aided by the decline of Dunwich, for in 1547 Edward VI granted the town its Charter of Incorporation. Like other towns along East Anglia's coast, Aldeburgh's fortunes were founded on the herring fishery, as well as its mercantile trade and boatbuilding, and like other communities on the Norfolk and Suffolk coast, over the centuries sea erosion has also taken a considerable toll on the town. With the North Sea to the east and the River Alde to the west, at one time there were three streets running parallel for nearly a mile in length. The town suffered severe erosion on several occasions, notably in 1591, when it was reputed to have lost approximately 20 ft of land in one day. The street to the east had been swept away by 1748, leaving the Tudor Moot Hall facing the open sea. Around eleven houses were also lost in a severe storm in 1779. The population declined and those who were left behind faced difficult times. Despite this, Aldeburgh was nevertheless described by the *Universal British Directory* as having:

> 'A commodious harbour for seamen and fishermen upon which account it is fully inhabited by these sort of people.'

Aldeburgh was noted by the anonymous *Universal British Directory* correspondent for the 'drying of red sprats, which are caught and dried in abundance, and exported to Holland'. Aldeburgh fishermen were also catching sole and lobsters. The writer recommended the White Lion as the principle inn in the town, and warned would-be visitors that, 'it is the only inn convenient for travellers', suggesting that those frequented by fishermen and sailors were to be avoided at all costs!

Aldeburgh fishermen, 1974. (Ernest Graystone)

By 1794, exporting sprats to the Lowlands must have been difficult. The French Revolution and the subsequent Napoleonic Wars had a damaging effect on the economy of the East Coast and of the country as a whole. The French had seized Rotterdam in 1795, which effectively killed off the export trade in cured herring and sprats to Holland. By 1801, Aldeburgh's population had declined to only 804. Fishing continued from Aldeburgh into the 19th century, however, with soles, lobsters, sprats and herring making up the catch in the

1870s. By the late 20th century, longshore boats were still being launched from the beach, a practice which had continued for centuries. In the 1960s, Aldeburgh was reputed to have had around a dozen boats fishing off its shores; by 1982 this had risen to twenty-five, with most registered at Ipswich. Today, however, that figure has declined to around five boats, most of which are now registered at Lowestoft.

Ipswich, the county town of Suffolk, has a history as a trading community and maritime heritage that dates back to its founding by the Danes. Despite its size, it had no fishing fleet comparable to those further up the coast at Lowestoft. Ipswich appears to all intents and purposes to be a port of registration only. Longshore boats can be found up and down the Suffolk coast from Southwold to Aldeburgh with Ipswich registrations (IH), but these are only small vessels compared to those once found at Lowestoft.

Ipswich was founded where the River Gipping meets the Orwell, a river which offered a natural harbour for ships. By the 13th century, the town was wealthy enough to apply for its charter from King John. King's Lynn and Ipswich were both involved with the wine trade and the Icelandic fishing. Merchant houses were built in the parish of St Clements, named after the patron saint of fishermen, which was close to the harbour.

By the end of the 18th century, Ipswich had problems with the Orwell silting up. Although trade declined, things had improved by the early 19th century, but by then any fishing, such as the Iceland voyages of previous centuries, had disappeared.

This did not mean that there were no fishermen working out of Ipswich. Quite the contrary; oysters were being caught on the Orwell until just before the Second World War. Doug Ellis of Ipswich, following his demob, recalls seeing members of the Crawford family fishing on the Orwell in 1947. He told me that Mr Crawford had a fishmonger's shop in Berners Street and the family had the fishing rights to part of the river. The fact they were using trawl nets suggests they were catching shellfish or eels. Like most fishermen, the sea was in their blood. Grandfather Crawford used to be a water bailiff at Ipswich. Once a year, Mr Crawford junior, although then in his 70s, would forego the River Orwell and load the family complete with fishing boat onto the railway and head off to Wales to fish for eels there instead!

Chapter 2

Fishing the Herring

'Coming into the port [Lowestoft] during the Herring Fishing was one hell of a sight … The Herring Dock was full of drifters, most of the time you were stem in. You could never get broadside there were so many'

Glen Chipperfield, former fisherman

The herring fishery had been part of East Anglian life since before the arrival of the Normans. At the time of the Domesday survey, herring was still being used as currency – as a due to the king, or to the local abbot, or lord of the manor in lieu of cash payment. Domesday also tells us of other ports that no longer exist; Gorleston, Kessingland and the city of Dunwich. Kessingland, for instance, paid Hugh de Mortfort a manorial rent of 22,000 herring – such totals give us a good idea of the size, not only of the fishing at the time, but of the community itself.

Casting the nets, LT 208 Rotha, *c.1938.*

Herring are migratory. They also stick to the same route year after year. They would swim past the Welsh coast up towards Scotland. Once reaching the Shetlands, they would then travel down the east coast of Scotland and England towards their breeding ground off the coast of France. Although the herring were 'fat' by the time they reached the west coast of Scotland, the fact that they were quite oily made them not too good for curing. However, by the time the shoals reached the area known as Smith's Knoll, off Great Yarmouth and Lowestoft, they would have reached their optimum condition for curing. It was for this reason that the two ports of Lowestoft and Yarmouth were set to haggle over the herring for centuries. The quality of the fish once it reached the coast off East Anglia also attracted fishermen from other parts of the country, as well as from the Netherlands, the German states and later the Scots.

Great Yarmouth received its charter in March 1209, giving the town its first powers of self-administration; and from that time on, the town sought to control the lucrative North Sea fishing and the mercantile trade to and from the city of Norwich – a trade which eventually made Yarmouth very rich indeed. Great Yarmouth's Free Fair of Herring was the greatest fish market on the East Anglian coast. It lasted from Michaelmas (29 September) to St Martin's Day (11 November). The Fair attracted fishermen from France, Flanders and Holland, as well as England. In turn, catches brought here were sold to merchants from England, Europe and even the Middle East.

In the 13th century, Yarmouth did not consider Lowestoft to be a rival. In fact, the main animosity it had was with the Cinque Ports of Kent, mainly because they insisted on taking part in the administration of the Free Fair. The hostility this caused during the 13th and early 14th centuries even led to the opposing fleet's fighting each other and the loss of at least 25 boats from Yarmouth. Tensions peaked in 1297, when Yarmouth and Kentish men settled their differences while escorting the Flemish king home! The Statute of Herrings was passed in 1357, which confirmed the Cinque Port's rights and clarified the extent of Yarmouth's control over the fishery.

As the Cinque Ports lost their power and their hold on Yarmouth dwindled, the town had to turn its attentions to a potentially more calamitous problem. Since late medieval times up to the present day, the sandbank on which Yarmouth is built had not only expanded southwards, but threatened to completely silt up the narrow channel leading into the River Yare and Breydon Water – all that is left of the entrance of the Great Estuary. This sandbank expansion had reached as far south as Corton by the early 14th century, some three miles north of Lowestoft. Yarmouth's attempts to secure an open channel for shipping were met with limited success until the early 17th century.

To the present day, Great Yarmouth has had seven havens. The most successful was the innovation of Dutch engineer, Johas Johnson, who created a new entrance between Yarmouth and Gorleston. Completed in 1613, of special interest was the curve at the entrance to the haven which not only built up the beach on the Gorleston side, but retained a stable entrance to the harbour which lasted for 400 years. In 1962, engineers renovating the old timbers had forgotten what had been learnt and did not take into consideration the deliberate curved shape of the entrance – the result was that Gorleston lost much of its beach and problems returned once more.

Lowestoft, in the meantime, began to profit by the continuous silting of the Yarmouth haven. Merchantmen wishing to evade the heavy duties they had to pay at Yarmouth, began unloading their goods, including fish, off the coast at Lowestoft. Yarmouth were clearly unhappy about this and successfully petitioned Edward III, in 1372, to forbid the loading or unloading of goods or the holding of a rival fair within seven leuks (leagues) of Yarmouth, unless at the port itself. Lowestoft was considered to be only five leuks from Yarmouth

Paddle tug United Services *coming into Yarmouth haven, c.1904, showing part of the 17th-century Johnson sea wall on the right.*

quay. A year later, several Lowestoft men were arrested selling herring within this seven leuk boundary. The north Suffolk town obviously did not take this incursion lightly. The power struggle between the two towns continued with various repeals and restorations. Yarmouth controlled the herring and, as Norwich was then the second largest city outside of London, eventually controlled the supply of fish and other merchandise inland, as well as the export of goods to the continent.

Edward III 'in his thirty-first year' according to Bloomfield's *History of Norfolk*, had decreed that 'the *hundred* of herring shall be counted by six score, and the *last* by ten thousand'. That was all very well, counting the fish was now regulated by law, but what was a *leuk*? Many considered it a league – the Roman league was approximately one and a half miles in length. Yarmouth said it was longer and claimed it measured approximately two miles. Others – mostly on the Lowestoft side – claimed that a leuk was equivalent to an English mile. The controversy would continue until the 17th century.

Lowestoft was now a town – the lord of the manor had applied for a market as early as 1308, believed to have been held on the Denes. In 1442, the town had its first purposc-built marketplace and it officially became a port in 1679. The final settling of the league/leuk argument – it was officially recognised as an English mile – did not ease tensions between the towns. Both Great Yarmouth and Lowestoft continued as rivals, if not enemies, until the 20th century – when by 1999, both towns had lost their fishing fleets and their livelihood, and both communities were in dire economic distress.

Some idea of the importance of the herring fishery can be gained by the number of boats fishing from the two ports in 1670, nine years before Lowestoft officially became a port. At this time Great Yarmouth had some 220 fishing boats as against Lowestoft's 25. It is also worth noting that although the fishermen of both towns tended to fall out with each other, sometimes leading to blows and the occasional bloodshed, many of the merchants whose ships sailed out of Yarmouth were also the very same men who had businesses – as well as owning fishing boats – in Lowestoft. In the 1790s, for example, Phillip Walker, the main partner in the Lowestoft China Factory was also a herring curer. He moored his ship, the *Phillip and Rebecca* at Yarmouth.

One of the earliest merchants to send Lowestoft boats into Scottish waters searching for herring was Daniel Peachey in the 1770s. The port was starting to feel the effects of the constant wars England had been having with the Continent, as well as the infant United States of America. Fishing was now in the doldrums – Lowestoft had found another interest becoming a health resort for the well-to-do, although it continued as a merchant port. Until 1781, when the Dutch wars

Lowestoft Herring Market, 1930s. (Paul Allison)

began to affect the fleet, an average of thirty-three boats fished from Lowestoft. A decline then set in which was only stopped with the Fishing Act of 1786 which granted a bounty for every herring caught, thereby alleviating any financial hardships and reinvigorating Lowestoft's ailing herring industry.

Chapter 3

The Two Rivals

For nearly 800 years, Great Yarmouth would remain the premier East Anglian port as it fought to control the herring fishery. This, however, would begin to change in the 19th century, as the port of Lowestoft increasingly began to dominate the East Anglian fishing industry.

The rivalry between the two towns had continued on and off for the best part of 500 years, although Great Yarmouth had always retained a fishing fleet to dwarf that of Lowestoft. In the early 1790s, Yarmouth boasted that it had the herring fishery off the coast to itself – or at least so it believed. The port retained 150 vessels just for herring fishing alone. One correspondent, an anonymous traveller visiting Yarmouth around 1793, assessed that during the season around 50,000 barrels of herring 'which some magnify to 40,000 lasts, containing 40,000,000 herring are generally taken and cured here in a year.'

As well as its fishing – which also included mackerel and 'north cod' – late 18th-century Yarmouth also retained an extensive mercantile trade. Unlike the Market Place in Norwich, where the fish merchants' and butchers stalls' were in separate areas or 'rows' away from the main market, there was no such division at Yarmouth's own great Market

Place, then the largest in the country, and still one of the biggest in Europe. The same 18th-century writer noted that not only were fish merchants scattered among the general stalls, butchers were also slaughtering cattle and sheep in full view of their customers in 'the centre of such an opulent town, resorted to by crowds of genteel company from almost every part of England.'

The Dutch had been severely hit by the Napoleonic Wars and this ultimately affected their trade, notably with Great Yarmouth. British blockades effectively destroyed the fishing monopoly that the Dutch once had, and although they returned to Yarmouth after 1815, not only had the Free Fair become a general market on the South Denes, but the Dutch eventually lost their monopoly in curing herring to the Scots, who were coming to Yarmouth in ever increasing numbers.

Yarmouth's harbour enabled the fishermen of Yarmouth to sail in bigger boats. The three-masted lugger was introduced at the beginning of the 19th century, and was much larger than the boats fishing out of Lowestoft, which until the early years of Samuel Morton Peto continued to be beached on the shingle in front of Lowestoft Denes. Vessels at Lowestoft had to be small enough for the crew to be able to drag them up onto the beach. Nevertheless, despite their size some of Yarmouth's own fishing boats continued to unload on the beach in front of the town walls, selling their catch at auction on the sands (also known as the Denes), just as they had done for centuries. This practice continued until the building of the Fish Wharf between 1867 and 1869, which gave Yarmouth a purpose-built covered fish market for the first time.

Although Great Yarmouth was without doubt the largest herring port on the East Coast, and with the bigger fleet, the East Anglian fishing port with the greatest development and fastest expansion was Lowestoft. The development of Lowestoft as a port created an increase in the number of vessels fishing off the East Coast. It was a development that would eventually influence the whole of the East

Herring landings at Great Yarmouth, c.1900.

Coast fishing industry for over a century. Lowestoft's first harbour was built by the Norwich and Lowestoft Navigation Company in 1830, linking the sea at Lowestoft to the city of Norwich via a canal and giving Lowestoft its first bridge. According to illustrations of the day, the canal connected Lake Lothing and Oulton Broad to the River Waveney via the New Cut and appears to have been used mainly by merchant vessels. Unfortunately, the project was not as successful as the original investors had hoped. At Lowestoft, the Navigation's lock gates suffered from the teredo worm, a seawater wood-boring mollusc, causing them to stick in the open position. It was from that point that Lake Lothing ceased to be a freshwater lake.

In the mid-19th century, Lowestoft had a reputation not only as an expanding fishing port, but for some 80 years as a health resort where the gentry and nobility came to take the air, bathe in the sea or

Off to the fishing grounds, Lowestoft, c.1899.

recuperate. After the Fishing Act of 1786, fishermen from outside the town, some from as far away as Ramsgate, Kent, began to move to Lowestoft and the town once more began to prosper. Even in the midst of the Napoleonic Wars, Lowestoft boats were reputed to have earned as much as £10,000 in six weeks.

In 1736, the vicar of Lowestoft, the Reverend John Tanner, made a survey of the town estimating the population to be around 2,200. The first official census in 1801 gave a population of 2,332 and by the 1811 census, its population had risen to 3,189. A year after the opening of the Norwich and Lowestoft Navigation in 1830, the population had reached 4,238. However, these increases were nothing compared to what was to come, and which was all down to one man – Samuel Morton Peto.

Born in Surrey in 1809, Samuel Morton Peto was an aspiring entrepreneur with an interest in the new railway craze then sweeping the country. Peto was living in Norwich where he was developing his idea for an Eastern Counties railway line from Norwich to Lowestoft. He had gone into partnership with his uncle, Henry Grissell, who already had an interest in the resort at Great Yarmouth. Peto decided to look towards Lowestoft, however, buying up the ailing Norwich and Lowestoft Navigation, as well as land on both sides of Lake Lothing. Part of his plan was to build a fish market connected to the railway when it arrived. Convincing the town's fishermen to use a purpose-built market was no easy task. Traditionally, local fishermen had landed and sold their fish in the open on Lowestoft Denes, just as their rivals were still doing at Yarmouth and elsewhere along the coast. For centuries, fishing boats had been hauled up onto the beach at the foot of the town on the cliff, but it was Peto's promise to both the fishermen and boat owners, made in Lowestoft's Town Chamber in 1843, that once the railway reached the town, Lowestoft fish landed in the morning would reach the Manchester markets first thing the following day, that convinced them to change their centuries-old habits. By comparison, acquiring land for a yet larger project at Lowestoft was a far easier matter.

In 1844, at his new home at Somerleyton Hall, Samuel Morton Peto asked the trustees of the town's lamplands, an area of undeveloped common land, how much they wanted for the sand-dune waste south of the Navigation known as Kirkley Heath. It is said that he offered £200, which they accepted with no qualms whatsoever – it was, after all, only a piece of wasteland. Peto, the legend continues, then invited the visitors who had just parted with their piece of unwanted property for what they considered the princely sum of £200, into his library where he promptly showed them what he proposed to do with this so-called useless land – a new health resort to rival Brighton; in effect, a new town that would go hand-in-hand with improvements to the Norwich and Lowestoft Navigation, including an enlarged harbour.

The railway eventually reached Lowestoft in 1847; the new Fish Market was built on the North Quay and from that moment on, the old rivalry between the two neighbouring ports entered a new, less hostile chapter.

Although the railway had arrived in Great Yarmouth in 1844, the Fish Wharf was only opened in 1869. The railway linking the Wharf with the main London line arrived in 1882, some thirteen years later. Lowestoft's own first purpose-built fish market – albeit a smaller structure – was constructed in the 1850s as part of Samuel Morton Peto's promise to the local fishermen. From the start, therefore, Lowestoft market had rail links both to the Midlands and to London. This fact alone ultimately influenced the expansion of its fishing fleet. Once the railway reached Lowestoft, the town, already growing, expanded even more rapidly. In 1851 the population stood at 6,580 and by the census of 1871, it had reached 13,620 and was still growing. Sir Morton Peto was declared bankrupt in 1861 but both the resort and port were now so successful that they continued to expand unhindered.

The Scots followed the fish to Lowestoft in the 1860s and their wives, sisters and girlfriends eventually followed them, packing the herring as the Dutch had done at Yarmouth decades before. From that point, Lowestoft became part of an annual friendly invasion that lasted until the end of the herring in the 1960s.

Meanwhile, the port was still expanding. The new Waveney Herring and Mackerel Market was opened in 1883 to deal with increasing numbers of drifters arriving at Lowestoft, not just from the two East Anglian ports but from ports as far afield as Ramsgate and Lerwick. In 1892, the Trawl Dock was expanded to the size we see today.

This was the era of what many called the 'Klondike', when from September to November each year, hundreds of sailing drifters crammed into the harbour for the Home Fishing, as it became known. Photographs of the time prove that, yes – you could cross the harbour without getting your feet wet!

The Waveney Herring and Mackerel Market, opened in 1883, seen here c.1910. (Paul Allison)

Trawlers had arrived at Lowestoft by the mid 19th century, although it was much later that local boat owners converted their own vessels to trawler-drifters to allow for all-year-round fishing. Trawling was very different to drifting. While drift nets lay fairly close to the surface, trawl nets disturbed the floor of the sea, taking up fish such as plaice and sole, and regrettably anything else that got in their way. Fortunately, at the time most owners remained dedicated to drifting. But whatever else was caught, it was always the herring that was the important staple. It was then, during the autumn Home Fishing that hunting for the 'silver darlings' reached its peak.

Crew of the smack Welcome *in Lowestoft Trawl Basin, c.1905.*

Each year between September and November, every nook and cranny in the two ports of Great Yarmouth and Lowestoft was given over to curing, pickling and packing herring. Each port in effect became what can only be described as huge fish processing factories. It was then that Gorleston and Southwold became vestiges of their old selves. From the mid 1850s, Gorleston became home to the Hewitts – a trawl fishing company that started in Barking, Essex. Fishing out of Barking was already on the decline when, in around 1856, Hewitt decided to move his family and his firm to Gorleston. By the early 1870s, Hewitts were among several boat builders and smack owners who were based at Gorleston instead of Great Yarmouth. Harrod's directory for 1873 lists fifteen smack owners (including Hewitt), at least four boat builders – and quite a few tailors! Reaching their peak in the 1880s, Hewitts had a fleet of around 170 vessels and employed 2,000 men. They also introduced the use of a dandy rig to Yarmouth boats instead of a lugsail, making for better manoeuvrability.

In 1878, the average age of a Yarmouth fishing boat was around ten to fifteen years old; several others appear to have been built in the early 1870s with some a few years earlier. More ancient vessels could still be found fishing, however. The oldest I discovered still working out of Yarmouth in 1878, was YH 432 *Hope* owned by C & W Pye. She was built in 1818 and weighed twenty-one tons unladen.

For a while, Yarmouth became the supreme trawler port, nevertheless it was not long before both Hewitts and Great Yarmouth had competition from Grimsby. The Hewitts company was wound up in 1901. A reminder of this once great fleet is the Tower House in Gorleston High Street, which was used as a lookout for any Hewitt vessels approaching the harbour.

Today, visitors to Southwold can be forgiven for having the impression that for much of its history this small town has been a pleasant but sleepy little seaside resort in the back of nowhere. Now the haunt of people retired from the film and media industry, this was once far from the case. Up until the first decade of the 20th century, Southwold had extensive fishing interests. Like Yarmouth, it was also an ancient herring port, and once was the seasonal base for the Dutch fishermen.

In the early 1890s, Southwold beach was crammed full of beachmen's huts, mostly used for storing fishing gear. The beach below the Promenade tended to be narrow and the huts, mostly built of tarred timber, were inclined to suffer from severe gales and were occasionally swept away by storms. Boats were brought up onto the shore by capstan, possibly using a similar principle that Lowestoft used for hauling its bathing huts out of the water. By the early years of the

A 1970s view of Southwold harbour. (Ernest Graystone)

20th century, there were around forty boats working off the coast at Southwold, with each allocated area worked by families. Their longshore boats were known as 'punts' and survived into the 1920s.

The Southwold port also played an important part in the autumn herring fishery. Scots fishergirls following the fleet down from the north had a base here, as well as at the two main East Anglian herring ports. I am told – but cannot substantiate it nor place a date – that many of these girls used to stay in the village of Walberswick on the opposite bank of the Blyth. The trade at Southwold was very successful, so much so that the town expanded its harbour to cope with the increased trade. There was one major problem, however. There were no direct road or rail links to its markets. Despite these setbacks, a new herring ring opened in 1906 and the new harbour a year later.

Regretfully, this success was not to last. Following the lean year of 1909, the number of vessels landing at Southwold, mainly Scots – Fifies and Zulus – declined. In the bonanza year of 1913, over 1,359,000 crans of herring were landed at Lowestoft and Great Yarmouth alone. When you consider that a cran of herring weighed 28 stones, that is 392 lbs, that means that over 38,052,000 stones – 532,728,000 lbs – of herring were landed at the two ports between September and December that year. However, whilst Lowestoft and Yarmouth revelled in their best ever season, fishing at Southwold had all but finished. Once on a major sea route, Southwold had been left

On board the smack Arizona *in 1908. (Brenda Leggett)*

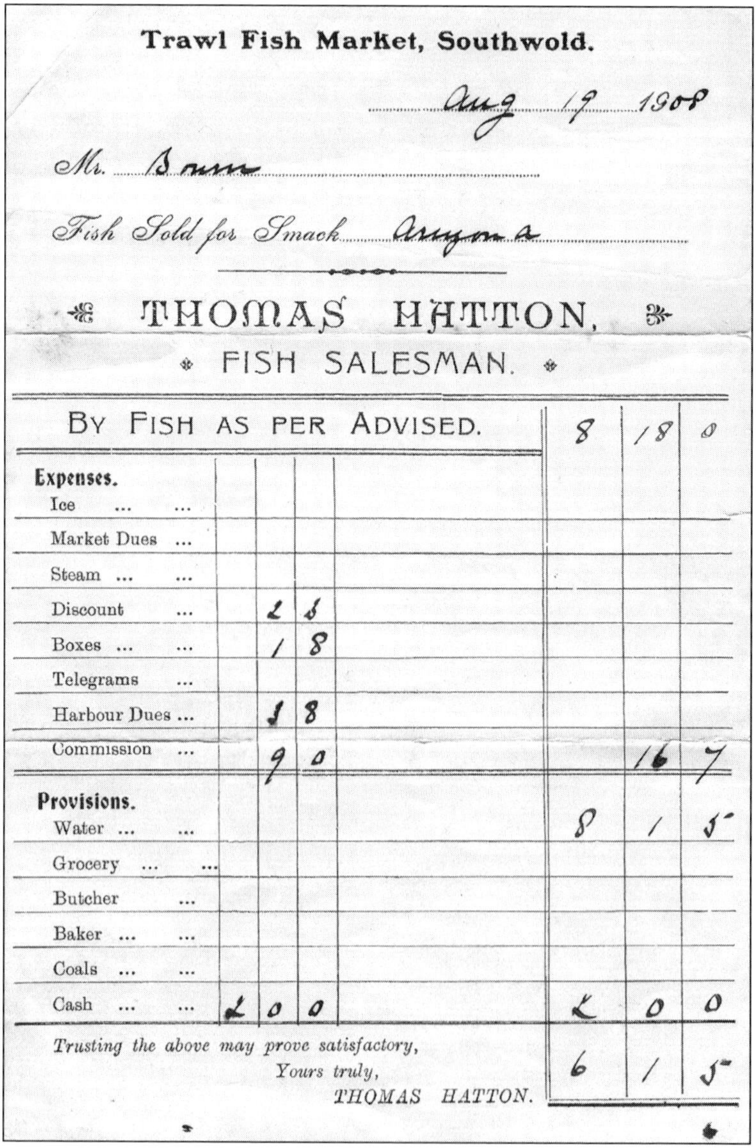

Trawl Fish Market, Southwold.

Aug 19 1908

Mr. Bonn

Fish Sold for Smack Arizona

THOMAS HATTON,

FISH SALESMAN.

By Fish as per Advised.				8	18	0
Expenses.						
Ice						
Market Dues ...						
Steam						
Discount ...		2	8			
Boxes		1	8			
Telegrams ...						
Harbour Dues ...		1	8			
Commission ...		9	0		16	7
Provisions.						
Water				8	1	5
Grocery						
Butcher ...						
Baker						
Coals						
Cash	2	0	0	2	0	0
				6	1	5

Trusting the above may prove satisfactory,

Yours truly,

THOMAS HATTON.

A 1908 handbill from the smack Arizona. *(Brenda Leggett)*

behind in the hustle and bustle of a new era – the only rail link it had with the outside world was the narrow-gauge Southwold to Halesworth railway opened in September 1879. Fish loaded at Southwold had to be transferred to main-line trucks at Halesworth.

Of those final days of Southwold's North Sea fishing little has survived as personal recollections. The Sailors' Reading Room at East Cliff, with its fine collection of old photographs is a good start, and occasionally rare pieces of ephemera still turn up, as in the case of Brenda Leggett, whose grandfather William Bourn was skipper of LT 297 *Arizona*, a sailing trawler that used to land at Southwold.

The Sailors' Reading Room at Southwold, c.1960s. (Ernest Graystone)

William had joined the *Arizona* in 1901, in what position his discharge papers do not say, however he was skipper in December 1908 when he left to join the *Iris*, owned by William Robbens. Like many on the trawlers based at Lowestoft, William Bourn came to the area from Ramsgate. Following his apprenticeship to James Wales, which, according to a list he wrote later of vessels he had sailed in from December 1877, this made him an apprentice at the tender age of nine. This also meant that he was second hand on R 388 *Excel*, a Ramsgate smack involved in the Home Fishing, at the age of fourteen, before he settled down to fishing off the East Anglian coast.

After the First World War, fishing at Southwold briefly prospered once more, when longshoremen began catching sprats from August to November each year. Regretfully, this was halted in 1939 by the outbreak of the Second World War. Today, as a small quiet resort, Southwold copes with a holiday trade and where Scots girls once used to stroll in their spare time having a 'guid auld blaither', cars vie for parking spaces. In 1901, at the height of Southwold's involvement with the herring, its population stood at 2,800. Not long after the Second World War, in 1949, its population remained in the region of 2,750; in 2010 it has declined to some 1,300 full-time residents. Longshore fishing still continues, of course. I am informed by local historian Sue Allen that only a handful of longshore boats now fish from Southwold. However, these add character to what is now a charming little coastal retreat.

Ernie Dewhurst used to fish from Fleetwood and later from Lowestoft; he then became a longshoreman working between Lowestoft and West Mersea:

> '[Although] we never land at Southwold – there are no facilities – we used to go in there sometimes. But of the five or six boats that work out of there, all the fish comes to Lowestoft because there's no [fish] market. How things are at the minute, they've settled for wet fish shops on the Harbour.'

Ernie also explains about the problems of landing at Aldeburgh,

> 'I used to work for a bloke, David Ward, and we used to go into Aldeburgh 'cos he used to live there. The thing with Aldeburgh is that it's a bugger of a place to get into 'cos there's a sandbar there, and you come up like a little river, so it's not a place you would actually fish from.'

Several local boat owners named their vessels after family members, the most famous being YH 89 *Lydia Eva*, built at King's Lynn for Harry Eastick. Hazel Bliss of Bungay told me that her two older sisters, Evelyn Joyce and Marjorie Grace, both had fishing boats

Steam drifters in Lowestoft Herring Basin, 1912.

named after them. Evelyn's was SN 198 *Evelyn Joyce*, built in 1912 at Great Yarmouth.

Until 1900, North Sea fishing was mainly done by sailing drifters or trawlers. Although steam-powered paddle tugs had been at Yarmouth and Lowestoft since the mid 19th century, they were used to tow the smacks out to sea on days when the weather was calm. Steam fishing vessels, mostly from Scotland and Tyneside, made their appearance in the 1880s. As late as the 1890s, Lowestoft boasted that it had 250 sailing drifters but only one steam drifter.

Steam made life easier for the crew, as men once needed to look after the sails could now help with the catch. The first local steam drifter was LT 718 *Consolation*, built by Chambers & Colby for the Kessingland-based boat owner, George Catchpole, in 1897. Despite being more expensive than a sailing smack, the advantages of steam were that it did not rely on the wind, it could hold more fish and it was faster than sail – an important factor when the sooner you got into port, the better the price you got for your catch. Once the other owners saw the advantages, steam power quickly caught on. By 1913, Lowestoft boasted 350 steam drifters and only a handful of sailing drifters. During the interwar years, practically every fishing boat built was powered by steam.

One such steamer was LT 167 *Hosanna*, a drifter-trawler built by Chambers at Oulton Broad, Lowestoft, for Albert Edward Beamish of Kirkley in 1930. Albert Beamish got his young daughter, Beryl Joan, to launch his new boat in 1930. Now in her nineties, Beryl still has fond memories of her father and the *Hosanna*:

> 'We had six boats … We had just steam, Dad didn't have anything to do with sailing boats. And Chambers, they built the *Hosanna* … I was eleven years old when I christened *Hosanna* and I can remember my mother grumbling, "You don't want to keep buying boats at your age." "Well, I think we'll just have this one – we'll call

One of the last steam drifters, LT 225 Kirkley*,*
seen here at Lowestoft in 1959.

> it *Hosanna*." I think years and years ago there'd been a smack connected with Dad in some way called the *Hosanna*, and I think that's how she got her name.'

Beryl remembers her father with affection and still recalls his meticulous dress and business habits,

> 'He had a little short beard, he was a very smart man, my father; [he] always carried an umbrella. And when he used to go down to see if they were looking after his boats alright, the people that were repairing them or anything, my goodness he was sharp! He went once and they'd put in a piece of wood he didn't like. He said, "That

On board the steam trawler Grosbeak, *February 1947.*

> must come out. I'll stay here until you take it out." They used to call my father the colonel, 'cos he was that sort of a man!'

The *Hosanna* proved her worth when she won the Prunier Trophy in 1938. She survived the Second World War and continued to fish up to 1975. Although she was converted to diesel in 1960 at Richards' shipyard in Lowestoft and fitted with a 335hp Rushton engine, Beryl recollects the boat was powered earlier by a Thornycroft petrol engine. It was 'a huge great big engine. She could do eleven knots at top speed.' Several steam-powered fishing boats survived into the 1950s before they were either scrapped or converted to diesel.

Glenn Chipperfield first went to sea in August 1954, straight from school onto a Lowestoft steam drifter. Wanting to go into the Merchant Navy, his father got him a berth on the Lowestoft boat *Neves* instead, which was then moored at Aberdeen,

> 'That first trip I was "spare hand", so you were "the boy". We had eleven crew; I was spare hand, so I got all the [rubbish] jobs. We were a coal drifter and my job was to trim [sic] coal. You took a cargo of coal on deck so all your bunkers were full … as the trip went on you had to shovel towards the bunker so that the chief engineer could continue shovelling the coal into the boiler. It was reasonable on a decent day because you could have your bunker lids open for air.'

Chapter 4

Trades and Industries

At one time, practically every industry in an East Anglian port would have had some connection with fishing. To every man employed at sea, it was said there was an estimated one hundred men and women employed on land in one sort of job or another to do with the fishing. In this chapter I have included a sample of the industries once found on the East Anglian coast. Some – like the smokehouse and the swill maker – went back centuries; others, such as the long-distance road delivery services, only started after the Second World War.

The Smokehouse

The two main ways to preserve fish were pickling and curing. The Dutch are believed to have discovered the principle of curing herring in 1386. From then until the Napoleonic Wars of the 1790s, they held the monopoly in cured fish. The method they used involved gutting the fish immediately after they were caught, then packing them into barrels with each layer covered with salt. By the 1830s, when the Dutch eventually returned to the fishing industry after the Napoleonic Wars, this trade had effectively died out, to be superseded by the Scots who adapted the same techniques for pickling herring, gutting the freshly caught fish and packing them in barrels between layers of salt, which proved so popular in Germany and Russia.

Kippers are believed to have been devised by John Woodger of Seahouses, Northumberland, in 1846. Woodger also had a smokehouse in Market Row, Great Yarmouth. Until then, both Lowestoft and Yarmouth had tended to produce red herring, a particular favourite with public houses. The reds took up to seven days to cure and were noticeably salty – hence the customers in the pubs that sold them were liable to drink more than usual!

Curing or smoking herring involves hanging the fish over a slowly burning fire of oak or ash tree chippings, but preferably oak. The fires would burn for up to four weeks. One reason why Lowestoft was originally split into two – with the town at the top of the cliff and the smokehouses at the foot – was the threat of fire, in an era when most buildings had thatched roofs. Until the arrival of the electric light, unprotected naked flames were everyday hazards in a town which consisted of a variety of timber, daub and wattle structures, alongside flint and brick buildings. Even well-constructed buildings, including merchant houses and workshops, were prone to the risk of fire. The walls of fish houses became saturated with fish oil which then seeped into every part of the building itself. If it then caught fire, and if the wind was in the wrong direction, as had happened at Lowestoft in March 1645, what began as a small fire could quickly turn into something far more terrifying, destroying life and property.

Smokehouses were once a common sight in both main herring ports, as well as in Gorleston and Southwold and King's Lynn. At one time it was believed that following the demolition of the North End

A tower on Yarmouth town walls, c.1905, once used as part of a smokehouse. (Ernie Childs)

The Old Smokehouse, Raglan Street, Lowestoft, the oldest surviving smokehouse in Lowestoft, seen here in 2000. (Author)

of King's Lynn in the 20th century, the smokehouses there had all been lost. However, the discovery of an invoice dated 1893, from Thomas Westwood, a fish and ice merchant of St Ann's Street, in the centre of the old community, hinted at the location of a smokehouse. This smokehouse was at the back of what is now True's Yard Museum in North Street. By chance, following the museum's renovation of an old workshop facing the two remaining cottages in True's Yard itself – the remnant of the last King's Lynn fishermen's yard communities – it was discovered that the smokehouse building was still there. True's Yard Museum which reopened in January 2010 now has the last surviving smokehouse in the town. How long it was working producing bloaters and smoked mackerel has as yet to be confirmed, but it is believed to have been in existence from at least the 1880s, if not earlier.

In Yarmouth, despite being for much of its history the largest herring port on the East Coast, if not in England, very few smokehouses survived compared with its neighbour, Lowestoft. The demise of the East Anglian herring fishery in the 1960s saw many fish curers go out of business, but unlike Yarmouth, some of Lowestoft's smaller smokehouses continue to function, if not flourish, curing other fish apart from herring. Smoked cod, haddock and mackerel can still be found at two of these smokehouses in the town. A few, such as the Anchor Fish Stores at the rear of Alexandra Road, Lowestoft, reopened to a later generation of cured and fresh fish lovers. Reynolds' smokehouse in Raglan Street, now known as the Old Smokehouse, is reputed to be the oldest in Lowestoft. It still offers cured as well as fresh fish in a building believed to date back to the 17th century.

Nevertheless, two notable smokehouses survive in Yarmouth. The extensive premises of William J Burton in Blackfriars Road continued as the Tower Curing Works until 1988 when it finally closed. It was originally built around 1850 and enlarged in 1880. As probably the last complete curing complex in the town, it was purchased in 1998 with a view to creating a museum dedicated to the town's fishing heritage. It reopened to great acclaim as the Time and Tide Museum in 2004. Its seven smokehouses are still intact and are now part of an important exhibit in the museum. While still a working smokehouse, the Tower Curing Works had everything needed to cure herring including its own cooper's workshop, salt store and, of course, smokehouses, which even today retain the aroma of oak smoke fires, a smell which fills the whole museum.

Not far away, on the opposite side of Blackfriars Road is the Old Smokehouse at Great Yarmouth Pottery whose entrance is in Trinity Place. Its uniqueness is its location as part of the hidden fishing heritage of Yarmouth, and the fact that as a smokehouse it dates back to the 18th century. Not only is it the oldest surviving smokehouse in the town, but it was constructed inside one of the town's 13th-century towers. Yarmouth's problem of being on a sandbank is that it has always had limited space in which to build or expand. This meant businesses sprang up in the most unlikely places. The advantage of the medieval tower was that as part of the town's defences it was relatively fireproof. It continued as a smokehouse until after the Second World War. The present owner and custodian is Ernie Childs. A gifted artist and potter, Ernie was born into a fishing family who lived in the old medieval rows in the town,

> 'There were sixty women working here. Upstairs there used to be one little office with one little phone in. I got a ledger going back, oh, I don't know … to the 18th century. I believe they did the Royal Navy, this place did.'

On the walls of the smokehouse museum are photographs of the fishergirls, then as elsewhere mainly from Scotland, preparing fish for curing. Although the building was only large enough for one

Fishergirls at Yarmouth, c.1930. (Ernie Childs)

smokehouse, the cramped working space did not deter the girls from their chatter and singing during the working day.

The herring would be soaked in brine prior to curing and then hung across the smokehouse chamber and left there for a few hours before being smoked over fires built up from oak chippings. Timber mills such as Jewson's used to produce wooden boxes for the fishing trade. They also had large quantities of oak sawdust and shavings which the curers eagerly took off their hands.

Alan Jones, whose father founded C H Jones, curer and wet fish shop in St Peter's Street, Lowestoft, recalls his younger days in his dad's smokehouse,

> '[The shavings] were ideal for smoking herring. If suitable sawdust was placed on the shavings to create smoke [they slowed] the rate of burning. The best sawdust by far came by truck from Watton [Norfolk] and also from Darby's of Beccles.'

Transfer to the smoking chamber was always done manually; even today men still climb inside a chamber to secure the speets to the interior. Some smokehouse chimneys could be as tall as forty-six feet.

C H Jones' smokehouses were on the premises. I quote from Alan Jones' autobiography, *Life at 174 St Peter's Street – a Lowestoft Childhood,* at the point where he is talking about preparing the fish for curing:

> '*After the draining period they would be tipped into the "trow" (as in row). I believe it should have been "trough" as in pigs' trough – but I guess trow sounded better. It was approximately three feet six inches long, two feet wide and about nine inches deep, with arms each end and grooves at the top to support the kipper baulks. The wooden baulks were about three feet long, one and a half inches wide and three quarters of an inch deep. There were eight pairs of hooks on each side, each baulk thus holding eight pairs of kippers. The hooks*

were known as tenterhooks. When speeting bloaters onto the speets, if the gill of the herring was broken, these ones were hooked onto tenterhooks and then known as "tenders". The kippers in the trow, once on the baulks, would be placed on a "horse" nearby, ready to be handed up into the smokehouse and care taken that a sufficient gap was left between the baulks to enable the smoke to circulate through them.'

Not all the herrings cured were caught off the East Anglian coast. Alan Jones again:

'The slack time was coming to an end. It was a case of waiting for the herring to swim around the top of Scotland and down the east side. I can't remember having herring from Lerwick, but I do know we had a regular supply from Fraserburgh. The herring would soon be feeding on a red substance and that meant they would shortly become very fat and boats would soon be working out of the port of North Shields.'

Alan also explains more about his father's smoking process for bloaters and confirmed the retention of the older system of counting herring:

'When smoking the bloaters, it was important to keep the oak log fire going as slowly as possible. The movement from the hot air could weaken the heads of the herring, causing them to drop. When speeting bloaters or red herring, twenty-two would be put on a speet and six speets made a hundred, or to be correct a long hundred: 132 … Two herring in each hand made one warp; thirty-three warps made a long hundred of 132; and one hundred long hundreds made 13,200, which was called a last.'

Butcher's smokehouse in the old Beach village at Lowestoft was probably the most easterly in the country. Mark Butcher believes the firm was started by his great-grandfather around 1900. After the last war, the business, which included a wholesale fish merchant at 14a Herring Market, was run by the three Butcher brothers, Peter (Mark's father), Ralph, and Bill. Bill ran the smokehouse which was situated on Gas Works Road. Mark Butcher remembers both the smokehouse and the market as a ten-year-old boy:

> 'The other two [Peter and Ralph] ran 14a Herring Market. I used to go down the smokehouse. I know they used vats of water – or maybe they were brine – they used to wash the filleted herring, pin [sic] them onto the roods and they [the men] used to be like a monkey, climbing up and hang them all. I think he had great sacks of oak shavings which he put into little pyramids and then he lit them. I don't know if he did bloaters or not.'

Scots fishergirls pickling herring on Lowestoft North Denes, c.1899.

Still in existence and not far from the Sea Wall, Mark estimated the height of the curing chamber to be no taller than 'an average semi-detached house'. Inside everything was black and sticky with decades of curing kippers. And of course there was the odour of the oak firings. 'He really did smell,' recalls Mark Butcher of his uncle. 'He never could get rid of that smell. He carried it wherever he went!'

Like those at Great Yarmouth, the smokehouse at Lowestoft also employed Scottish fishergirls, but by the late 1960s the only people left were Bill Butcher, a chap called 'Fatty' – so called because he was as thin as a rake – and a casual hand recruited from the two other brothers' business on the Fish Market. Large-scale curing has now gone, however, as have the days when the aroma from smokehouse fires hung over Yarmouth and Lowestoft like a pair of heavy curtains! Some very small smokehouses do survive including, so I'm told by Kevin Barrett, an ex-fish lumper himself, several in the Pakefield area. One or two also maintain the more ancient tradition of wind-drying fish.

Mike Moyes and Brian Blowers remember two such fishermen, Mike first,

> 'In Stradbrooke Road [Pakefield] he got his own smokehouse and sellin' kippers. He got it in his garden … and they're all done the proper way. They're nice if they're all done properly, like smoked haddock – you go in the shops and they're all dyed. You get a pan full of dye.' 'I always remember Billy Carew [sic] used to wind dry herring. He used to hang them in his garden, thread them through on a pole through the gills and leave them out to dry.'

The pervading smell of fish must have made an impression on the author, Charles Dickens, when he visited Yarmouth in 1849. He stayed at the Royal Hotel, on Waterloo Road and set parts of his book, *David Copperfield,* published between 1849 and 1850, in the area. This is how he describes Yarmouth in *David Copperfield*:

'When we got into the street (which was strange enough to me), and smelt the fish, and pitch, and oakum, and tar, and saw the sailors walking about, and the carts jangling up and down over the stones, I felt I had done so busy a place an injustice.'

The Scots fishergirls

One of the greatest losses to the East Anglian fishing industry has been the disappearance not only of the Scots drifters, but of the fishergirls themselves. Starting at Lerwick in the Shetland Isles, these girls moved south following the fishing, travelling from port to port, gutting, pickling and packing herring by the million as they went. Bedecked in oilskin skirts and rubber-boots, in East Anglia vast numbers of these Scots lassies could be found at Great Yarmouth, Gorleston and Lowestoft, and, at one time, at Southwold, throughout the autumn herring season. In the early years of the 20th century some

Scots herring drifters at Great Yarmouth, 1930s. (Peter Killby)

of these girls were as young as thirteen – in a number of cases even as young as twelve. Several of the married lassies would bring their bairns – their young children – with them.

Of course not all fishergirls were Scottish, some were local, but it is the Scots that are best remembered. This friendly invasion initially made its appearance in the two main ports following the increasing numbers of Scottish boats in the 1860s. These girls had the reputation of being some of the hardiest workers in the industry. Their nimble fingers were ideal for gutting, pickling and packing the herring into the thousands of barrels once seen stacked everywhere during the herring season. Many came down by train year after year, mostly from the Islands and some from other parts of Scotland, working in the open in the bitterest weather with their hands wrapped in binding to protect them from the salt and brine throughout the cold winter nights, not to mention to protect themselves from their extremely sharp knives. What fascinated East Anglian folk was their language – mainly Gaelic – which has been described as a cross between singing and talking; and their hands – which were always on the move. During a break, to keep their hands nimble the girls would be forever knitting – knitting while sitting, knitting when walking across the market, knitting while strolling in the town. Even when work was finished, their hands were never idle.

At the height of the Home Fishing, fishergirls at Great Yarmouth and Lowestoft would gut sixty herring a minute – that is, one every second. They would work in teams of three – two girls gutted the fish while the third packed them into the barrels. This procedure would be repeated wherever they went.

By the 1930s, the girls also came down by bus, as well as by chartered trains. Landladies taking in the girls as lodgers at Lowestoft, Gorleston and Great Yarmouth went through an annual ritual of stripping the room bare, leaving the minimum of furniture and covering walls and furniture with either newspaper or brown paper.

Gutting herring at Denmark Road, Lowestoft in the 1920s.

Box rooms – or small bedrooms capable of taking only one person during the summer – would now accommodate three girls in a single bed. Not quite as draconian as it sounds – the girls usually worked in shifts and took turns sleeping. The stripping bare of the room was a necessary precaution because of the fish oil and brine which got onto the girls' clothes and into their skin. Footwear was usually left outside the house and their belongings would be packed in a box and stored under the bed. If more space could be found in the house, then that was also rented out for the season – even the front room was sometimes put into use, however small!

Ernie Childs of the Great Yarmouth Pottery and Smokehouse, in recalling his early life on the Fish Wharf in the 1950s, retains his affection for these lassies,

> 'I've grown up with these old girls, y'see. They were "girls" no matter how old they were. When you're down with them, it's lovely. [As a young child] you used to be among them, and they were singing away. It was a wonderful atmosphere. It's a wonderful way to grow up. You couldn't understand half of them, though! I used to spend most of my time down the Harbour.'

Ernie also recalled the hundreds of boats moored at Yarmouth during the height of the herring season in the 1950s. On a Sunday morning when all the Scots crews had gone to church, locals wishing to visit relatives on the other side of the river would climb across from one

Here the girls are packing pickled herring at Great Yarmouth, c.1905. (Ernie Childs)

boat to the other to get from Great Yarmouth to Gorleston, rather than walk the half mile around and over the Haven Bridge!

In the 1920s, the Scots girls would arrive at Gorleston at the Beach railway station. One company, Waters & Son from Wick, Caithness, had their curing facilities opposite Gorleston lifeboat station. As in Lowestoft and Great Yarmouth, pickling plots were located where any space could be found, including between Pier Plain and Bells Marsh Road.

The swill maker Frank Philpot, in recalling the fishing industry, confirmed that despite their incredible speed of one fish gutted per second, very rarely would you see evidence of any emergency first aid!

Although young at the time, Maureen Fryers of Lowestoft clearly recalls these later Scots girls:

> 'We just couldn't believe it when we saw them walking down the street knitting away, talking, and doing all this lovely "Fair Isle". And they were just jabbering away. But didn't they work hard. They were really hardy women.'

If youngsters were not playing among the barrels they watched the barrel makers – the coopers – at work, as John Fryers, born in 1941, the son of a fisherman, also confirms:

> 'When we were kids, we used to play on the barrels on the Denes. One of the stalls in Whapload Road was the barrel maker … we used to stand there watching him make the barrels and putting the hops on. He used to love having an audience. The problem was when the fish died out, so did the coopers. All the trades went with it.'

Jim Nolloth, a fisherman, also recalls playing among the barrels:

Scots fishergirls on Lowestoft North Denes in the 1920s.

> 'When you go down Mariners Score, if you remember that used to be all the pickling plots; thousands and thousands of barrels, all where Birds Eye is now. All round the Lavender Laundry … even along towards Christchurch, almost towards the Trawl Market, right along to the coastguards. In my boyhood days we used to play at the bottom of Mariners Score.'

But for those youngsters whose mothers opened up their homes to the girls, even the children themselves were enthralled over the dexterity of these strangely dressed women and their ever-moving fingers – a fascination that has now passed into fishing legend. Jeanette Nolloth, among many, remembers not their speech, but again, primarily their knitting!

> 'They were lovely. They used to knit us Fair Isle jumpers, Tammy hats, gloves, they used to do. They were so quick how they packed them herring in those barrels.'

Many of the girls also worked in fish yards located in the town itself. Iris James lived in Sandringham Road, Lowestoft, close to one such yard. She was born at the end house, near to Roman Hill Secondary Boys School. Before the Second World War, White and Willows' yard located between Sandringham Road and Avondale Road would have been very busy pickling and packing herring. Wartime and the construction of air raid shelters then took the place of laughing fishergirls. Iris' family was bombed out along with my great-aunt, who lived opposite her at 57 Avondale Road.

Both Lowestoft and Great Yarmouth took the Scots to their hearts. In 1905, Messers H Le Grice, haberdashery, drapery and children's outfitters at 114 and 116 London Road North, offered Lowestoft tweed for 1*s* 0½*d* a yard. It is very much to my regret that I do not recall the Scots lassies, despite the fact that they were in Lowestoft up

Packing the final layer of pickled herring, Lowestoft, in the 1930s. (Paul Allison)

to the end of the herring fishery in the 1960s. I put this down to the fact that my father was also a Scot, he spoke very broad Aberdonian. As a wee bairn, Scots dialects tended to be an everyday occurrence around me, so much so it seems, I took them for granted, and more the pity!

Rarely do we see in surviving photographs just how hard these girls worked, or the tough unpleasant conditions they sometimes endured. Underneath all their gear and weather protection, however, many of the lassies were quite pretty. Needless to say several romances occurred and many stayed to marry into East Anglian families. The fishermen, too, occasionally found themselves wives along the Norfolk and Suffolk coast.

The only Scots involvement I remember from my childhood occurred around 1955. I was a small boy standing on the Trawl Dock with my parents and younger brother watching the Scots boats going to sea. One was approaching the pier head when it got stuck on the sandbank inside the harbour mouth. I distinctly recall a crew member at the bow shouting to one of the other Scots boats to throw him a line to get off the bank.

Whatever I missed when I was a child, in my early working life I do recall at least one fish curer employing a number of women in their processing sheds, and that was F H Phillips & Co, later known as Marinpro Ltd. The business was founded in the 1880s in Newcastle upon Tyne, by Frank Henry Phillips, who also had considerable fishing interests elsewhere. Their premises were in Norwich Road, Lowestoft, and remained there until the 1970s. Paul Hoey joined the company in 1963 as a management trainee,

> 'By this time the herring fishing was well into decline, plus eating habits had changed. FHP at that time had an office at 25 Herring Market, plus the yard at the end of Norwich Road. In those days the herring was sold at auction in the herring ring. Although there was an auction, each class of buyer had their own minimum price, as marinators our minimum price was 77*s* a cran. In those days there was a television programme called *77 Sunset Strip*, so every time we bid 77 all the other buyers would sing out *77 Sunset Strip*! After the auction, a cup of tea in the Europa café on the market, we would then tip the herring into kits from the ships' boxes for transportation to Norwich Road.'

Once at Norwich Road, the process began to turn herring into rollmops and Bismarcks, the latter a delicacy popular on the continent. Even then, mechanisation was already creeping into the industry.

> 'We had several filleting machines … they were round and pulled the herring around the machine by the tail. The gut and roe were sent down a wooden flume, the fillets collected in a basket.'

FH Phillips' yard, Norwich Road, Lowestoft, December 1953. (Paul Hoey)

The fillets were then put in pickle. The recipe was a secret known only to one employee, 'Ducks' Knights, and the required amount poured into *killers*, which were shaped like open-top barrels,

> 'We then tipped a basket of herring fillets into each killer, gave it a good stir and left it for four days … we then packed the herring into wooden barrels with a weaker pickle and salt. The cooper put the head in the barrel, then into our chill store went the Bismarck herring.'

Most coopers had gone by the 1960s; however, Phillips maintained their own barrel-maker until the end.

> 'The wooden barrels were made by our own coopers during the non-fishing seasons … When I first worked, the cooper was Sharmy Muse [sic]. He was a local man and always wore a trilby. [When] he died he was replaced by two coopers who came from Peterhead. Stanley Finey lived on site in a caravan, George Black lived in Gorleston and came by bike every day!'

The Yarmouth Swills

As their name suggests, these baskets are unique to Great Yarmouth, to be found nowhere else in the country. They are believed to have been introduced by the Dutch, and unlike baskets found elsewhere, swills have no corners or hard edges to bruise the fish. Just why they are such an unusual shape has been lost in time, but one suggestion is that it enabled herring to be scooped more easily out of a ship's hold without doing too much damage to the fish themselves. Despite its

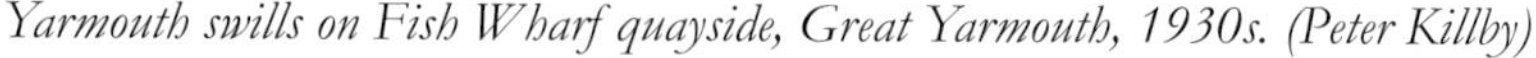

Yarmouth swills on Fish Wharf quayside, Great Yarmouth, 1930s. (Peter Killby)

antiquated shape, the swill remained in use well into the 20th century and only disappeared with the demise of the herring fishery. Swill making will soon be a lost craft; there is only one man left from the days when manufacturers produced baskets of all descriptions for the industry, not just for Great Yarmouth but also for Lowestoft and as far afield as Fraserburgh.

Frank Philpot left school at the age of fifteen in 1948 and went straight into basket-makers, Stanley Bird's at Friars Lane. Getting any kind of job was tough in those years immediately after the war. Young school-leavers had to vie with servicemen reconstructing their lives as they returned from the war. Youngsters, therefore, were grateful for any work they could get. This particular job, I suspect, was initially to tide Frank over until he found one he liked. However, fate would dictate otherwise,

> 'It was a bit of a shock when I went into it. I wanted to be a bricklayer, like my dad, but unfortunately he died very young. I couldn't get a job with the company that promised me a job, so I went down Estcourt Road in Yarmouth to get [me] a ticket. "We'll send you down to Stanley Bird's for basket making." "Basket-making? What the hell am I doin' down there?" Anyway, I went down there and talked to the guv'nor, Stanley Bird, and he say "we'll give you a trial". It lasted forty-eight years.'

Frank's apprenticeship took three years. His first wage was sixty shillings and the first job he was given was to make sausage hampers. Each took approximately an hour and a half to make,

> 'I looked at the foreman and said "sausage hampers?" They were little old things and they held about two pounds of sausages, and they wanted hundreds of the damned things. Six by nine inches bottoms and then stake them up – they were terrible.'

The young Frank Philpot (left) and friend, with Stanley Bird's lorry. (Frank Philpot)

Having started off with basic basket-making, Frank eventually found himself working with several of the old traditional swill makers at Bird's, many then in their seventies. Some of the men he worked with as a young apprentice had been making baskets and swills for over thirty years, some for as long as forty years. 'They were good boys,' he recalls.

Frank's mentor was Harry Keen, who started his working life at fellow basket-makers Henry Roberts in the Rows. Harry's own introduction to swill making was reputed to be when his father gave him to Roberts in exchange for half-a-dozen rabbits; this appears to have been in the early 1930s, a period when both the Lowestoft and

Great Yarmouth fishing was in the midst of the Depression. Frank started off making quarter-cran baskets which held around 250 herring, the subsequent swill would hold anything up to 1,200,

> 'I spent several years just on those alone. Harry Keen taught me to make swills. It was pretty hard goin', I can tell ya … You started off making the swill with a bale with twisted willow on top … After making the figure of eight, you had your main ribs, bow ribs – about eight different sorts of ribs.'

Eight hazel ribs were also used in the process,

> 'When I'd learned how to make them I used to make four a day. They were very sturdy. None of the Scottish ports had them, up and down the coast nobody ever ordered any – it was purely a Yarmouth swill.'

Frank considered that the best willow for making swills was red willow – which was also the hardiest. Usually, strands of ordinary willow were used. The hardest part was cutting the whips – the branches – even when soaked. In the days before he could afford his own car, Frank would cycle home, his hands so sore and split with cutting and working with willow stems that he could hardly grip the brake levers to stop.

Bird's building in Friars Lane was three storeys high. There were twelve men on the third floor, while the girls worked on the ground floor. Stanley Bird's was a family firm that had been making baskets for over a hundred years. Founded by Stanley's grandfather, they began in the 1860s, in an era when there were at least six basket-makers in the town; however, even by the standards of the time, the 1950s and 1960s, conditions at Bird's remained fairly Dickensian – there was no such thing as health and safety and no one got paid when they took their annual holiday.

Some of Stanley Bird's basket-makers, Great Yarmouth, 1950s. (Frank Philpot)

Occasionally the younger employees would go out and gather the willows:

> 'There was ten acres of willow at Burgh Castle. The site is still there, but not as a willow field. You got paid 1s 6d a bundle.'

A trick used by the youngsters was to place grass reeds in among the stems – needless to say that was quickly found out! Later Frank also worked at Bird's School Road works at Runham Vauxhall.

At the age of 21, Frank learned to drive, and being the only one who could drive, he used to go down to the Fish Wharf with twenty to thirty swills at a time on the back of Bird's lorry. When the autumn herring finished, Frank was busy repairing swills for the following season. These would be stored in the workshop's cold, rat-infested bays,

> 'You could hear the rats in among the old ribbings. When you came to the last [swill], they came out just like that!'

General repairs allowed the swills to last for another season. Quarter-cran baskets were also made for Lowestoft Fish Market. A traditional design, as were the swills, these baskets could still be found in use up to the late 1990s.

All the years Frank worked at Bird's there were only four other companies that made swills – three in Great Yarmouth and one, surprisingly, in Lowestoft. An old-established firm, Newton Robert Daines had been basket-makers in Duke's Head Street since before 1901. In the 1930s, as N R Daines & Sons, they also had a basket works in Battery Green Road, opposite the Fish Market. By 1952, Daines' two sons were based in Florence Road, Pakefield. 'They were a couple of nice chaps, they were. They didn't do a great deal of 'em, but they did a good job,' says Frank.

Back at Great Yarmouth, each swill had to be officially checked – a governmental weights and measures inspector would come down to make sure all the swills met the required standards. Frank remembers these officials well! 'They were nasty buggers, they were!' One of the tests they used was to try and pull the swills apart. They also checked the quarter-cran baskets which had to be just right and had to hold 250 herrings. Every basket and swill was stamped to show where it had been made; YH for Great Yarmouth, LT for Lowestoft, plus the year of manufacture.

Making swills and quarter-cran baskets took up to six months of the year. Each workman had his own individual mark which was also useful when on piecework. Although the work was strenuous – humping long strands of willow up three floors before it could be worked, Frank nevertheless looks back on his time at Bird's with pride.

Following the collapse of the fishing at Great Yarmouth, basket-making had to find new markets – the unusual shape of the swill became a log basket. There was also a half-sized, scaled-down version. As much as he enjoyed his work, at the age of forty-nine, Frank branched out on his own, expanding into general basket-ware and selling to the coast's tourist trade. He also continued to make fish baskets.

After retiring from Bird's in the early 1980s, Frank retained his fondness for working with willow. Eventually he opened the Basket Shop in Wrentham, Suffolk, seven miles from Lowestoft, which would later be taken over by his son, Stuart. 'It's a very, very beautiful old shop. Wrentham is a superb little village.'

Everybody there knew Frank as 'the basket-maker'. And as the basket-maker of Wrentham, not only did he attract attention from Southwold and Walberswick, but also attended fairs, including several at American air bases and travelled as far afield as Hatfield House. One of his occasional, unusual orders was for a hot-air balloon basket!

However, it was the death-knell of a once great trade. Stewart Philpot remembers those final days well; 'It was rather sad to see all this history disappearing making little knick-knacks for the home,' he recalls sadly.

Although swill making in Yarmouth has ceased, all is not lost for the future, however. Examples of Frank's swills can still be seen at Great Yarmouth's Time and Tide Museum, as can some of his tools.

Chapter 5

The 20th-century Fishing Industry

'You could actually get the smell in the air of herring … My grandfather, because he was a lumper, we used to live on fish. My grandfather's cat used to live on prime whiting! I'm fish orientated – I could eat fish seven days a week'

Jim Nolloth

Although boats and equipment improved over the decades, life for the fisherman was never easy. Generally speaking, everything was done manually and almost everything required muscle power. Only in the latter end of the 19th century were there any signs of mechanical innovations, such as the steam capstan, for instance. Invented by Elliott & Garrood, engineers and boilermakers

XXVIII. Advertisement—Beccles

Elliott & Garrood,

LIMITED.

Engineers, Boiler Makers,

IRON AND BRASS FOUNDERS,

Ingate Iron Works,

BECCLES

Branches—LOWESTOFT, GT. YARMOUTH, BRIXHAM, NEWLYN, BUCKIE and FRASERBURGH.

Specialities :

The 'Beccles' Patent Steam Capstans

High-class Propelling Machinery,

Boilers, Donkey-Pumps,

Oxy-Acetylene Weldings, etc., etc.

All Inquiries will receive prompt and careful attention.

Telegraphic Address: **"ELLIOTT GARROOD, BECCLES."**

BEWARE OF IMITATIONS.

Give us a call at Beccles, Gt. Yarmouth, Lowestoft, Fraserburgh, or Buckie, when requiring anything in the Engineering Line.

Trade advertisement for Elliott & Garrood's steam capstan, Beccles, c.1919. (Hazel Bliss)

at their Ingate iron foundry in Beccles, nine miles from Lowestoft – the appropriately named 'Beccles' patent steam capstan was quickly adopted by boat owners. However, as we've seen, other innovations such as the steam drifter took longer to catch on. Engineers and shipbuilders including Richards and Brooke & Co, both at Lowestoft, and Pertwee & Back and Crabtree & Co at Great Yarmouth, not only built boats, but would eventually construct the engines that powered them.

Although sailing smacks had gone by the end of the 1930s, steam-powered drifters and trawlers survived until the 1950s. The first diesel drifter made its appearance in the 1930s; but it was only from the 1950s onward that they came into their own.

In the 20th century, fishing boats improved beyond recognition.

The wreck of the Sparkling Nellie*, Lowestoft, 1903.*

Not only did steam power, petrol and diesel take over from sail, enclosed wheelhouses replaced the open tillers seen on sailing smacks. Post-war designs also made for safer vessels to work in. Echo sounding and radar, and by the last two decades of the 20th century, satellite, saw skippers sitting in wheelhouses more akin to aircraft than fishing boats! Despite all these innovations, life at sea remained hazardous. Although some things had changed for the better, one object remained unalterable, and that was the sea.

The North Sea has always been a fickle mistress – one moment calm, gentle, serene and at ease; the next vicious, angry and dangerous – a living, deadly wall of foaming water threatening to lift a boat up and swallow it whole or to fall on it with a vengeance and smash it to pieces. It was at times like these that the skill of a crew was needed to get a vessel back to port. And sometimes only just – as an anonymous writer recalled in a letter once to be found in the late Jack Rose's collection,

> 'It would be about 1923 or 1924 that there were severe gales off Lowestoft and many fishing smacks were overdue. I remember on one occasion on my way to the railway station by tram in the company of another lad to catch the 8.30 train to Beccles, when we saw a small crowd gathering on the quay opposite.'

Playing truant to see what was happening, the two anonymous schoolboys joined the crowd,

> 'We saw a smack which had obviously been damaged in bad weather coming through the piers. Its mast was broken and a temporary sail had been rigged up. As the boat neared the quay we could see that members of the crew were lying in the gunwales and had tied themselves there. A man was also tied to the tiller. All were obviously in a state of collapse.'

The boat eventually crashed into the quayside,

> 'The owner, a Mr Catchpole, who wore a large trilby - a short man, but big built - was there and as he saw what was happening, called for someone to call the ambulance, taxis and a doctor.'

The smack had been seven weeks overdue and was thought to be lost with all hands. Members of the families hurried down to the quayside and Mr Catchpole, himself, was overcome with emotion. The anonymous writer continues,

> 'Men rushed onto the boat and carried the men one by one to the bow where Mr Catchpole carried each one to a waiting taxi or ambulance. As he brought the first man in his arms [onto the quayside], he started to sing the hymn *Nearer My God to Thee* and the crowd took it up, a most touching scene which I shall never forget.'

Glen Chipperfield went to sea from August 1954 to December 1960, and recalls one experience he had on the *Shepherd Lad*,

> 'I got washed one side of the deck to the other. Ernie, who was a friend of my dad's, was a good skipper but a bit wild. We finished trawling and we were hell-bent in getting to market, and of course he swung it [the jib] round and I was knocked [from one side to the other]. We were still on deck 'cos you put your batten down, put a cover over to make sure everything was water-tight.'

A fisherman for forty years, Nelson Chipperfield, Glen's father, had an even luckier escape when he was washed overboard and by a miracle washed back again. Occasionally crews were not so lucky; sometimes

LT 335 Loch Lorgen aground at Great Yarmouth, with the Caister lifeboat on standby, December 1963.

both men and boat would disappear to be seen no more.

Traditionally, fishermen wore clothing that gave some indication as to which port they sailed from. This continued into the first decade of the 20th century. Not just clothing to protect against the elements when working on deck, if a crew member got washed overboard and his body was washed up later on the coast, the type of garment would give a clue to the deceased's home port. A Lowestoft fisherman would wear a tan jumper with a black silk wrapper around his neck and boots with pointed toes. A Pakefield fisherman wore a red handkerchief, a light tan jumper, duffel trousers, and Wellington boots. A Kessingland fisherman could be identified by his long slop in bright tan, a peaked cap and Wellington boots. A blue jumper and duffel trousers with a slit at the bottom was the style worn by Great Yarmouth fishermen. Scottish fishermen would wear a blue calico top and jersey with pearl buttons on the shoulder. Even the weave of the jersey (known

Charles Wilson, cook on the smack Research, *wearing a traditional fisherman's gansey, c.1914. Wilson was killed in June 1916 when a German submarine fired on the* Research.

generally as a gansey) or jumper would further clarify what port a fisherman sailed from, and possibly who he was. If they were anything like those made for the King's Lynn fishermen, these ganseys were made from a five-ply oiled wool. The weave or texture had to be tight to allow seawater to run off. One thing all ganseys had in common was that they were knitted in the round like socks – they had no seams.

On shore, however, dress code was altogether different! In the late 1950s, at least one of the East Anglian ports, Lowestoft, could boast its own unique style of fisherman's casual suit – bell-bottomed or flared trousers with an individually cut jacket to match. The colours the young fishermen chose – lilacs, pinks, lavender, bright purple, etc., showed the confidence they had – and, of course, most importantly, it attracted the girls. These quayside dandies – you wouldn't have dared call them that to their faces – have only recently been recognised as part of the town's fishing heritage. The young men generally frequented two tailors in the town – Hepworths in the town centre on the corner of the Marina and London Road North, and Lawrence Green in the High Street. Both made individually styled suits to order which matched the flamboyant nature and confidence of these young seamen. Needless to say, this was also the Teddy Boy era when a number of Saturday night fights took place in the town.

Tom Field, who sailed for Boston Deep Sea Fisheries, used to get his suits made in Aberdeen in the 1950s, 'First thing on shore after being paid … was to be measured up for a nice new suit.'

The craze then started at Lowestoft – blue suits, black suits and so on. The different colours were strictly a Lowestoft innovation. While in port, Tom would occasionally pop up to Lawrence Green and stock up with shirts and ties. Unfortunately, as Tom also recollects,

> 'When the Teddy Boys came down, there was always a lot of trouble. Once you started on a fisherboy you got the whole fleet after you!'

Glen Chipperfield remembers his suits well. As soon as he arrived home from a voyage, it was straight round to Lawrence Green's in the High Street to get measured up for a suit,

> 'We bucked the system 'cos it was the Teddy Boy era with drainpipe trousers and such like, and we used to have as wider bottoms as you could get! I got married in a suit with twenty-two or twenty-three inch bottoms. All bright colours, with pleats down the back.'

Seamen from the Villages

Of the many people I interviewed for this book, several came from the East Anglian countryside, not just from villages close to the ports in question, but from areas not traditionally associated with the fishing industry. One reason for this move towards the ports was, of course, the seasonal laying off of casual farm labourers after the harvest; farmhands would then look for the most part towards Great Yarmouth and Lowestoft for work. For much of the 20th century, particularly in the heyday of the herring, this also meant that boats were never short of potential crew members.

This movement towards the coast was a tradition that went back centuries. Men tended to seek out work wherever they could find it, despite any hardship or risks. Across the two counties of Norfolk and Suffolk, fishermen came from as far inland as Mettingham and Bungay; as well as, of course, villages nearer at hand. Historian Rita Flatt confirms that even in such a small community as Lound in north-east Suffolk, to make ends meet a number of families there tended to be involved with the autumn fishing. Lound is approximately five miles from Yarmouth; but despite this, there was always a move towards the boats at Lowestoft. This might have been influenced by the fact that Lound is on the periphery of the Somerleyton estate, or the fact that the nearest railway station was also at Somerleyton.

LT 1197 Lord Carnarvon. *Len Brighton, 4th from the left standing in the back row, and Walter Brighton, the first from the right seated, came from Mettingham, near Bungay, as did two other members of the crew. (Hazel Bliss)*

Hazel Bliss, of Bungay, also confirms the affinity of the land with the coast. As well as working on the land or travelling to Lowestoft or Yarmouth to go drifting, Bungay had the advantage of another source of employment during the hard winter months,

> 'If you didn't go fishing, you went on the Maltings. The Maltings were owned by a brewery in Burton on Trent.'

Traditionally there had always been fishing off the coast at Winterton and at Hemsby, small villages north of Great Yarmouth. Needless to say, villages on the Norfolk side of the Bure and the River Waveney influenced many to travel to Yarmouth. In the 1970s, I came across a retired photographer, Harold White, then living at Filby, a small community just outside Yarmouth, who was adamant that before the war, men from the village working on the boats at Yarmouth used to walk from Filby to the Wharf, some seven or so miles away. This being said – some crossed the county line – Ernest 'Jumbo' Fiske, one of the greatest Lowestoft skippers in the post-war years came from Kirby Cane, in Norfolk.

Travelling to the coast was made easier by the arrival of the railway – Somerleyton is a case in point. Until the 1960s and the dreaded 'Beeching Pill' when many branch lines were closed, most small communities were never far away from a station; although not all, for example, Blundeston and Lound on the Norfolk/Suffolk border. Even after the arrival of the railway, a walk of five miles or more was nothing to anyone looking for work.

Although Blundeston, the village made famous by Charles Dickens in *David Copperfield*, was only three miles from Lowestoft, the author chose to set parts of his book in the rival town of Yarmouth. However, it appears that most villagers from Blundeston would have travelled to the closer port of Lowestoft to work on the boats. Dickens' preference to Yarmouth over Lowestoft was despite meeting with Samuel Morton Peto, who invited the author to stay with him at Somerleyton Hall.

Kessingland lies on the Suffolk coast just south of Lowestoft. By the end of the 19th century, little of the old fishing village remained. The arrival of the London toll road in the 1780s to the west meant that it had become more of an inland village. The 'Sea Row', a once thickly populated area of the older community, by 1854 had been swept away by the sea. Kessingland's fishing fleet had already long disappeared and what few boats remained tended to fish out of Lowestoft. As with other ancient communities on the East Anglian coast, encroachment by the sea has always been a problem. Although

Kessingland beach in the early 20th century.

Kelly's directory for 1927 suggested that this erosion had ceased and that Kessingland beach was then building up, this optimism proved false; by 1930 the beach in front of the village was again disappearing. There was no protection from the sea following the storm of January 1937, when many cliff-top buildings were lost. A year later the sea finally broke through, reaching as far inland as Latimer Dam, crossing the A12 London Road, and severing the Great Yarmouth to London road.

The instability of the cliffs did not stop the villagers' involvement with the North Sea fishing, however. Several Kessingland families retained their boat-owning interests as late as 1927 and beyond, where in two areas in the village alone – Church Road and the Beach – one could find the Catchpole family – Arthur, Arthur George and Edward; G.E. Curtis; Arthur Gouldby; Harold Hervey; James Keable; George Manthorpe; Benjamin and Lewis Strowger; the Thackers, Samuel and Samuel Junior, and William. This does not mean we should forget the

Tripps, the Uttings, or Alfred Beamish. And all in a village with a population estimated in 1921 to be 2,056 – just less than the population of Lowestoft at the beginning of the 19th century.

Life in port and around the market

In the post-war years, fishermen were usually in port for forty-eight hours after a ten-day trip, before they returned to sea. One of the first things the average young crewmember did, once setting foot on land, was to go for a drink and talk about the success or otherwise of their latest trip. Each boat had its favourite haunt and certain pubs became notorious – the examples of the Spread Eagle and the Stone Cottage in Lowestoft were repeated elsewhere in Great Yarmouth, King's Lynn and at Ipswich. It was a tradition going back hundreds of years, as we've already seen at Aldeburgh.

Of course, drinking as soon as you got on shore wasn't just restricted to the home ports. Glen Chipperfield remembers one such occasion on the French coast,

> 'I've seen Jumbo Fiske [flat] on the deck of the *Sarah Hide*, either in Boulogne or Calais, I can't remember which, with the union jack over him, ****** as a newt! The French thought he was dead!'

But like all fishermen who met this six foot six inch tall man (and to many it seemed almost as broad) Jumbo was highly respected along the East Anglian coast. 'He was a great person; an amazing bloke – a good skipper,' recalls Glen.

Of course, not all countries were as understanding about the fisherman's tipple. Tom Field was a crew member of the *Boston Swift*, which with another Boston boat, was sent across the North Atlantic to Nova Scotia, Canada, in October 1954. The journey was comparatively uneventful apart from the second vessel's engines breaking down. The *Swift* had to tow her for the rest of the way into

Ernest 'Jumbo' Fiske, skipper of the Suffolk Warrior *and winner of the Prunier Trophy in 1964. (Ernest Graystone)*

port. Naturally the first thing the crew wanted after such a long crossing was a drink! No bar could be found so they bought their beer from the Canadian equivalent of an off-licence. Wandering back to the ship they were arrested by the local police – not because they were drunk or on Canadian territory without going through the usual customs procedures – but because they were drinking in public, which was an offence. Their return journey back to Britain on the RMS *Samaria* was quiet by comparison!

Returning to life at sea, John Fryers, in recalling his father's time as a fisherman in the interwar years, stated that Fryers senior tended to sum it up in one word – 'Hard'. He fished out of Great Yarmouth and Lowestoft on steam trawlers travelling around the coast,

> '[He] fished for twenty-odd years and then his mate got drowned at sea – chap by the name of Sammy King. In fact I worked with Sammy King's brother, Fred, when I was at Ballys, although I didn't know it at the time. My father fished from Lowestoft, Great Yarmouth and fished out of Scotland – North Sea and deep-sea fishing – but not what we would term Icelandic. I think he worked for Small's.'

Jim Nolloth agreed:

> 'You're out there, working hard; I've been on the deck for that man [the skipper] for thirty-six to forty-eight hours on a stretch and nearly got washed overboard. That would be the *Montserrat* then, when she shipped a hell of a sea – filled the thing right up.'

John Fryers' reference to *Bally* was to the Bally shoe company, one of many industries then in Lowestoft.

Even when safely on land, and especially when times were difficult, life tended to deliver a few hard knocks, particularly in the depression of the 1930s, when trawling was on the decline. For some this meant only one thing – the workhouse. At Lowestoft, this was the Mutford and Lothingland Poor Law Institution in Union Lane, Oulton. John Fryers again,

> 'You got no alternative; if you didn't work, you didn't get fed. My father used to turn up at the Workhouse at Oulton everyday for a

> day's work, like the majority of people from Lowestoft who were then out of work.'

Maureen Fryers' father, Harry Cone, was a lumper, one of those men who turned up at the quayside in the early hours of the morning in fair weather or foul to unload the catch from the boats ready for the auction,

> 'He used to go down to see the board outside [the fish market] to see what was coming in. If there was nothing on the board there was no work. If there were some coming in, there'd be the names of the boats. He would go in probably around eight at night. Usually the lumpers went home when they were finished. He didn't. He then went filleting. He was good – he taught me. He always had work. If there were no boats in, he always did filleting. He was paid a straight wage. He used to come home after work with fish-bones in his nails and I used to pull them out and they'd all be festering. Afterwards he used to dip them in blue disinfectant. My father only went once on the boats and that was to Lerwick for two months. I believe he was fishing for cod.'

There was no unloading the catch for auction on a Sunday; this would start at approximately one minute after midnight on Monday morning.

In the 1970s, lumpers generally started on the stroke of midnight. In Lowestoft, the Europa Canteen on the Herring Market was open at all hours, even as early as two or three o'clock in the morning, to serve the needs of both fishermen and lumpers. There were a few long tables at which the men were served steaming hot tea and whatever was on the menu. I was told many years ago that a previous manageress used to send mugs of tea and coffee whizzing down the tables. I used the canteen once or twice in 1962. It was always busy but I had the impression that once you put your foot through that door, you inevitably stepped back in time; you could feel the ghosts of previous generations of fishermen and dock workers looking over your shoulder!

Although there is no denying that life at sea was tough and occasionally extremely dangerous, some youngsters, nevertheless, travelled considerable distances to be on the boats. Reggie Brine was a prime example. Born in Kent in 1942, he had spent his holidays in Suffolk as a schoolboy after the war. The sea always fascinated him and the first chance he got, he returned to Lowestoft as a teenager. His first job on board was as a deckie (a junior deckhand):

> 'Immediately I left school I signed on the *Ada Kirby* in December 1957. I was fifteen. The *Ada Kirby* was a wooden boat.'

Unlike many of his East Anglian peers, Reggie relished going to sea and preferred the wooden hulled pre-war drifters still in use to the more modern metal hulls,

> 'I thought it was exciting. It was hard to begin with, but once you got used to it, it come easy. It was a hard life even once you got used to it all. I started as a deckie-learner, I done cook-learner then I went back on deck and finished up as bo'sun on the Lowestoft trawlers. I sailed as a deckhand on the *Wilson Line* after she was converted from a herring drifter to a trawler. My last boat out of Lowestoft was the *Chudleigh* under 'Polly' Wilson. I've fished out of Fleetwood, Milford Haven, but not Yarmouth. I remember filleting fish on deck during a trip and cooking it over a coal stove – luv'ly! We used to eat cold fish at night in between a couple of slices of bread and butter.'

Once back on shore and with his pay packet in his pocket, Reggie would either nip off to the bookies, or like most seamen after a rough trip, head for their regular public house. One thing all prospective fishermen have to contend with is their first trip. Reggie Brine took to being a fisherman right from the start; however, that was not the case with everyone. Even the best fisherman had to get used to the movement of the ship. Cooks occasionally preparing something on board other than the old standby of dumplings did not help. Jim Nolloth remembers his first trip in the early 1950s,

> 'I done one trip with Walter Spore … he used to be the skipper of the *W.F. Cockrell.* He took me to see if I'd like it. I stood on the deck on the back and I know now that they did 'easy meals'. When the cook did 'easy meals' he was doing cabbage. This cabbage smell was coming out of the galley and I stood on the rail. Of course I just heaved! You can imagine what's in you until there is no more! You're got a crackin' headache and you know jolly well you can't get off 'cos you're going out for eleven days! [The cook] he said, 'Are you havin' anything to eat, boy?' Anyhow, you get used to that after a while. I made the grade and he [Walter] took me on as a deckie learner.'

Employment in the industry was such that if you couldn't get the job you wanted, you made do in some aspect of fishing until something better turned up or until you eventually found that job you really wanted. This is what happened at Great Yarmouth and Lowestoft – but in the latter port, while Great Yarmouth's fishing was dying, certainly in the 1960s and into the 1970s it was still fairly easy at Lowestoft to go from one fish merchant to another, from one job into another, whether selling, filleting, lumping, or packing. If you were a hard worker and good at what you did, especially filleting, there was always work.

Maureen Fryers' mother worked in her father's fish-house in Clapham Road, Lowestoft, at the back of the Adrian Works. Born in

Mending drift nets, Great Yarmouth, 1930s. (Ernie Childs)

1912, as soon as she reached 14 her first job with her father was as a gutter. Hazel Arnold never worked in the industry, but her sister did – at least for a while,

> 'She hated it! She wanted to work in a shop but all she could get was a job as a filleter. My mother and my aunt were already there. She used to work down on the Denes. Eventually she got a job in a shop in Oxford Road, Lowestoft, before moving to London as a housemaid.'

Another vital industry was net-mending. My grandmother, Gertrude Coleman (née Capps) was a beatster, an old term for someone who repairs or mends – in this case drift nets. Hazel Arnold's grandmother mended nets at home. After the last war, many a housewife helped

East Anglian Fishermen's Training Association net-making demonstration, c.1970s. (Ernest Graystone)

with the domestic economy by repairing nets at home, sometimes in the open in the backyard or garden, or occasionally, as my mother did during the winter months, in the kitchen or living room in front of a roaring fire.

Most boat owners had their own net stores – Albert Beamish for example. Beryl Read (née Beamish) remembers,

> 'Dad had thirteen girls in the shed mending the nets. He used to take nets round to women to do at home … One of our men used to take Judy, my pony, and pick up the nets and bring them home [when they were finished]. We had two floors … a small room and a large room. A place for a fire and next to it they had a hook for

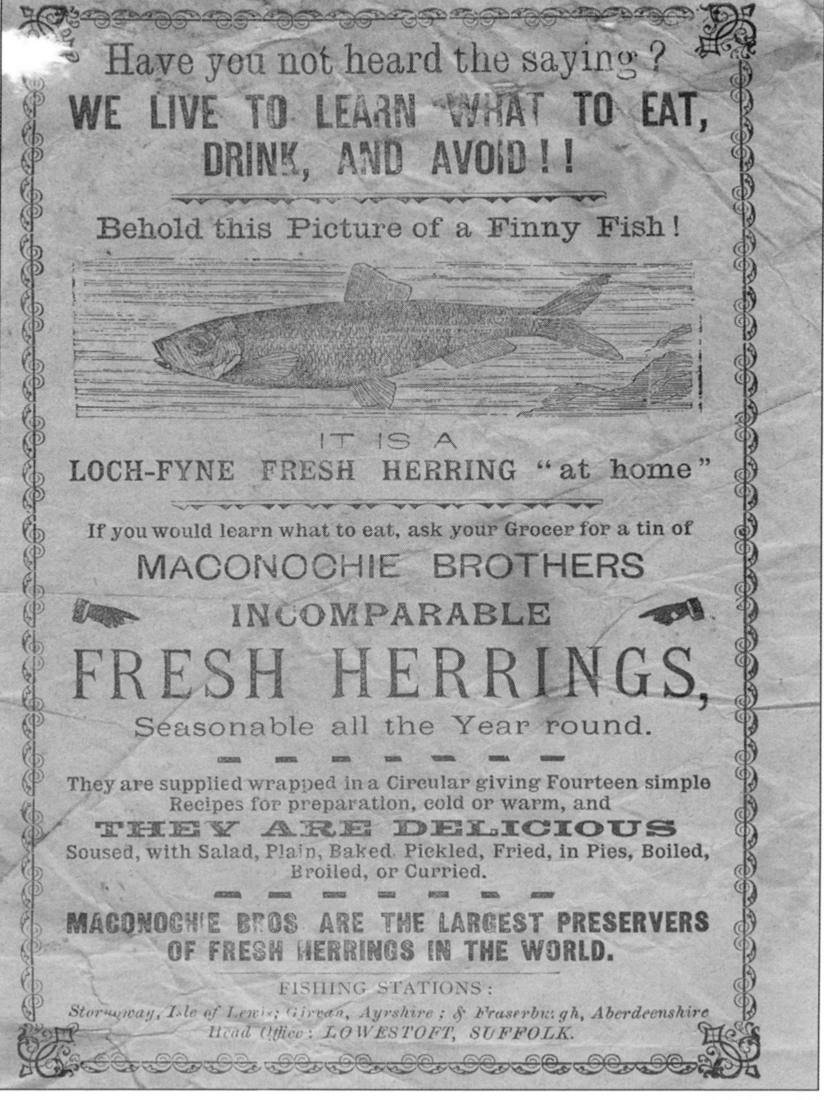

Handbill for Maconochie Bros herring, c.1900; the company was founded at Lowestoft in the 1870s.

> the nets. Some of the nets were badly damaged, others just had a tear or two and in half an hour they could repair it. Another one could take three [hours] or so or longer.'

Although not as busy as in pre-war days, the 1950s still saw both ports crammed with vessels. Maureen Fryers remembers these latter years recalling,

> 'There were 132 boats in the Trawl Basin at Christmas. This was one of the rare times after the war you could cross the Trawl Basin without getting your feet wet.'

Six years after the loss of the herring, Yarmouth's changing fortunes were unmistakable. By 1971, census returns showed that the population had dropped to 50,236 from a total of 52,970 ten years previously when the town was still a major herring port. Lowestoft, in the meantime, began to concentrate on whitefish and changed all its North Sea fleet to trawlers. The main catch was now plaice and, as such, Lowestoft continued to expand. The population had reached 52,267, and with an active fleet and almost full employment, Lowestoft had become the larger of the two fishing ports.

This continuing development of Lowestoft as East Anglia's main fishing port also saw the first large-scale fish processing factory on the Trawl Market. Food processing had been with us since the 19th century; two of the three notable factories included C & E Morton and Maconochie Brothers, both in south Lowestoft, maintaining the Scottish link in the port. Maconochie's, founded in 1870, became one of the earliest CWS food factories in 1929. The famous CWS *Jenny* tinned herrings were produced here. The third, Birds Eye, arrived in the late 1940s and saw the rise of extensive fish and vegetable freezer factories in both Great Yarmouth and Lowestoft.

Road verses rail

Until the 1950s, fish delivery inland continued to be by rail, whether to Billingsgate, the Midlands, or elsewhere in the country. Up until then, lorries were used to deliver fish to railway trucks either at sidings along the Fish Wharf at Great Yarmouth, or queuing along Commercial Road, at Lowestoft, waiting to load up at the goods yard at the Central Railway Station.

LT 295 Suffolk Maid, *a typical locally-built Small & Co. drifter of the early 1960s. (Ernest Graystone)*

Lowestoft had a problem, however. The railway station was one side of London Road North – which was also the main thoroughfare – while the fish docks were on the opposite side. This also meant regular traffic disruptions when fish trucks were shunted across from one side to the other.

After the Second World War, this reliance on rail changed. The cost of sending fish by train had become expensive. Wholesalers and boat owners started to look for an alternative way of delivering fish to their customers. Steam and petrol lorries had been used to cart fish to the pickling plots and smokehouses in the decades before the war. The

Icing up fish boxes prior to loading into insulated containers, Explorator, 1950s.

end of hostilities in 1945 saw a plentiful supply of now unwanted military vehicles. This resulted in several entrepreneurial souls looking towards road transport as a way of distributing fish direct to their customers' doors as fresh as it was on the morning it was landed. Up until then fish had been delivered to local shops on flat, open-sided vehicles. This was ideal during the cooler months of autumn and winter, but keeping fish fresh for long periods in summer was a considerable problem. Although ice was used to keep the fish cool, at the height of summer this wouldn't work for long, and tended to limit the delivery range to within a few miles of the market. It was in Lowestoft that Stanley Stevens, whose father founded Larvin & Company in the 1920s, had the idea of overnight deliveries using lorries with specially-constructed insulated containers. This eventually changed the way both Great Yarmouth and especially Lowestoft fish were delivered inland. This new system of fish delivery from the Fish Market by road, straight to the customer, succeeded in ending the monopoly of fish delivery by rail.

Of course such a pioneering idea needed financing. That was where Small and Company, who had their headquarters in Lowestoft at Waveney Road, facing the Trawl Market, came in. Small's was originally founded by Captain Thomas Small in the 19th century – but it was thanks to Fredrick Spashett, who joined Thomas Small as an office clerk around 1880, that the company became a guiding light and a major participant in the fishing industry at Lowestoft. Under Spashett, among other things the company became an extensive agent for the export of pickled herrings to Europe. Small and Company's house flag, a letter *S* between two bars, was the Spashett emblem and not Captain Small's as many once believed; it could be seen in one form or another until 1989. Small's had agents in most of the main British ports including Aberdeen, Milford Haven and Padstow in Cornwall. Stanley Stevens had an ally in a direct descent of Fredrick

Spashett, his grandson Tony Cartwright who had only just joined the firm when Fredrick died in 1945.

Thus, with help from Small and Company, one of the largest boat owners in the port, a new company, Explorator, was formed in 1947 and from that date a new era in the fishing industry was born.

As befitting a new company in the immediate post-war days, their initial fleet consisted of ex-military lorries, mainly Chevrolets. The insulated containers were, if I remember correctly, built in Lowestoft by P W Watson. At least I recall Watson's overhauling them in their workshop in Thurston Road. Those ex-wartime Chevrolets could be noisy, but the vehicle I best remember as a young child was their one and only six-wheeler, which was the noisiest vehicle I had ever been in and which had retained many of its wartime fixtures and fittings, including the observation hatch on the top of the cab. My father, Frank 'Jock' Robb, was one of their first drivers. Before then, he had worked for A E Balls, a haulage contractor based in Water Lane, Lowestoft, since his demob from the army in 1946. Despite the increase in money now coming in to feed a young family, my mother was not happy. It meant hubby would be spending several nights away on the road. And who knew *what* lorry drivers got up to!

The new company concentrated on the Midlands and southern England and to ensure as much of the area could be covered, Explorator opened depots at Wokingham in Berkshire, followed by Rugby, and in one of the old haunts much favoured by Lowestoft fishermen, Milford Haven. They also opened sales offices in Worthing and London. My father would spend weeks away at Milford, much to the annoyance of my mother, who eventually had two young boys to raise.

The lorries travelled overnight, sometimes in convoys, in all weathers to a tight deadline on main roads that until the early 1960s were little more than narrow country lanes. Very rarely were there any hold-ups, the occasional heavy snow being the exception. Later, the introduction of two-man crews eased the pressure – one drove while the other did all the lifting; nevertheless, the demands of delivering on time could lead to accidents, but compared with the sea itself, road transport was relatively safe. Even then, there were tragedies. One, I vividly recall, happened at Blythburgh, Suffolk, in the 1960s, when a fish lorry returning home to Lowestoft crashed into the White Hart, a solidly-built, 17th-century public house located at the bottom of a short but steep hill. The driver had fallen asleep at the wheel and hit the corner of the building, demolishing part of the wall. The lorry's cab almost disintegrated. The driver died on his way to hospital.

Chapter 6

The Beginning of the End

> 'Walking down the Market, there'd be fish from one end to the other. They'd start early and if the merchants didn't want it, it went for fish meal. The same with the herring – unsold it went for fish meal.'
>
> *Albert Stone*

And Jim Nolloth recalls,

> '[Walter Spore] used to work the fishing grounds at Smith's Knoll. He was what I call a skipper who used to work the 'smooth' out there. They work an area of ground on the seabed and they catch the flatfish. You didn't catch many soles, not wi' him; you used to catch plaice. Round Smith's Knoll, you always caught plaice. I used to think to myself, 'Crikey, you're gettin' it for nothing … Never did I think they would have to manage it when there weren't enough.

'When you go with a skipper who worked the 'smooth' you didn't have a lot of net mending. You steam out there to the fishing grounds, around three hours to the Smith's Knoll, or something like that, and then you'd fish. Depending on the fishing grounds, you might do a two-hour tow then you'd haul it aboard.'

There had been concerns of over-fishing and threats to the fishermen's livelihood as early as 1912, when proposals for trawling for herring were met with anxiety for the future from all the drifter ports, and not just in East Anglia. At least drifting only caught the herring shoals when they rose to the surface to feed, trawling meant that nets were dragged along the seabed, scooping up anything and everything, mature or otherwise. The *Yarmouth Mercury* reported on the fishermen's concerns over the proposals for herring trawling. There was a meeting at Yarmouth Town Hall between members of the fishing industry and local Members of Parliament. Alarm was also caused by the closeness of the trawl mess which allowed almost nothing to escape. Bloomfield's, a Great Yarmouth boat owner who also used trawlers, went as far as accusing trawlers from Grimsby of almost destroying the Scottish west coast herring grounds. At the end of the meeting, a petition was sent to Parliament pleading that 'action may be taken without delay to prohibit the use of this special trawl, or any other similar net.'

There was no evidence of a decline of herring stocks the following year, however; it was the best year on record for landings at Yarmouth and Lowestoft. In that bonanza season of 1913, there were 264 English boats and 742 Scots boats fishing out of Yarmouth whose catches totalled over 824,000 crans! And that was just one port. Lowestoft had just as much success with 350 English boats and 420 Scots boats landing almost 535,000 crans. A cran was twenty-eight stones. Eighty-seven per cent of the total catch landed went for export.

The outbreak of the First World War meant that there was no 1914 season; had there been, the herring would have disappeared a lot sooner than the 1960s. Although the shoals returned as usual in 1919 and 1920, they were not as plentiful as in pre-war days.

Demand for herring had also begun to fall. As late as April 1921, there were worries over the numbers of barrels of pickled herring still left unsold from the 1920 season on the Fish Docks at Yarmouth and on the Denes at Lowestoft. This state of affairs had been caused by political instability in Europe and the collapse of old pre-war markets. Both Russia and Germany were in the midst of economic collapse. Markets would pick up as the decade worn on, when pickled herring were once again exported to Germany, Poland and Russia, but in East Anglia, and with the recession of the 1930s, it was clear that although Lowestoft maintained 135 drifters and 84 drifter-trawlers, the resulting

Scots drifters unloading their catch at Great Yarmouth in the 1930s. (Peter Killby)

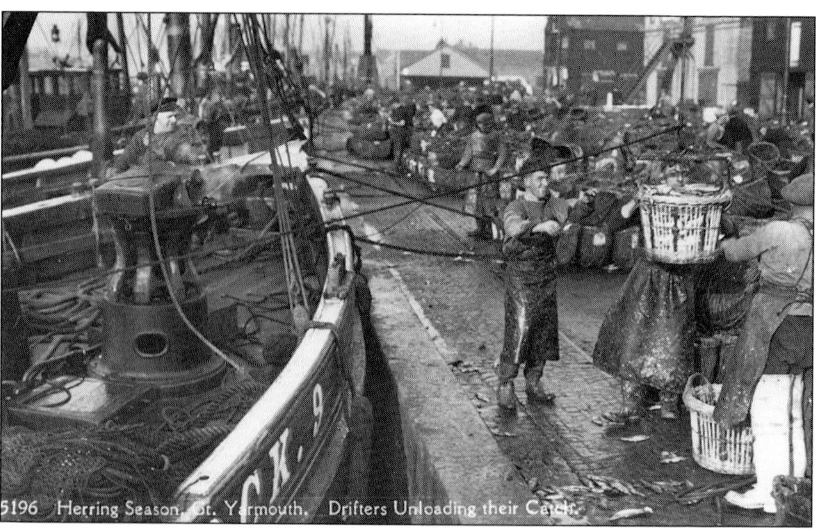

drop in the home demand for herring had caused Yarmouth's once great fleet to decline to 87 drifters and only 17 drifter/trawlers. Enter Madame Simone Prunier.

The idea of a Prunier Trophy came about when Madame Prunier, owner of the famous Prunier restaurant in London was approached following a discussion in early 1936 with Warner Allen, a friend of fellow restaurateur, CW Berry, about the plight of the British fishing industry. Madame Prunier readily acknowledged the herring as the 'King of Fish', but by the mid 1930s people were eating fewer herring than in previous years. Part of the problem was that it was a very bony fish. Madame Prunier understood that the herring fishery gave a livelihood to thousands of men and women, not just on the East Coast but also in Scotland – at which point she devised a competition for the boat that brought back the largest amount of herring in a single night, either to Great Yarmouth or Lowestoft. In this, she was aided by George Atkinson, Lowestoft's fisheries inspector. The trophy was designed by Charles Sykes. The first winner was Joseph Mair skippering BF 592 *Boy Andrew*, which landed 231 crans at Yarmouth. He and his crew travelled to Prunier's restaurant in London to receive the prize. In 1938, skipper William Bowles won the Prunier Trophy in LT 167 *Hosanna* for owners Albert Beamish, as Beryl Read recalls,

> 'The *Hosanna* got the Prunier Trophy catching the most fish for the season [sic]. We went to the Odeon Cinema and we were presented with the trophy. I didn't go on stage but Dad did – it was a lovely trophy.'

Despite the declaration of war, a trophy was awarded for the short 1939 season. Fishing stopped throughout the Second World War, however it recommenced in 1945. The Prunier Trophy returned in 1946.

LT 367 Dauntless Star, *seen with the Prunier pennant on its rear mast, 1965. (Ernest Graystone)*

For a period, the Home Fishing returned to something like its former self. In 1957, for instance, 111 boats fished out of Yarmouth catching 72,000 crans of herring. Despite the inevitable warning signs, however, herring continued to be chased round the coast of Britain and even as far as their breeding grounds off Cap Gris Nez. In hindsight, that was not a good idea. At Lowestoft, Small & Co. phased out herring catching in 1966, the last year the Prunier Trophy was awarded. The remnants of the drifter fleet made one final attempt in 1970 when only five Scots boats fished off the East Anglian coast. All they landed was sixty crans. King Herring was no more.

Lowestoft Fish Market in the 1970s. (Ernest Graystone)

Following the end of the herring fishery, the remaining drifters in Lowestoft were either sold or converted to trawling. Some had already been built as drifter-trawlers to enable all-year round fishing, but the North Sea is not very deep and trawling did nothing to help the situation. In fact it made it worse. Lessons had not been learnt by the loss of the herring.

As it was, Lowestoft now became East Anglia's premier fishing port. The failure of the herring fishery meant that Lowestoft's centuries' old rival, Great Yarmouth, ceased as a viable fishing port. Despite the Hewitt influence in the late 19th century, Yarmouth never changed over to trawling to the extent that Lowestoft did. The result was that the Norfolk port now saw its once great fishing fleet decline to little more than a handful of longshore boats. Following nearly two decades

Lumpers sorting and grading fish at Lowestoft Fish Market in the 1970s. (Ernest Graystone)

of intensive trawling, by the early 1980s it was clear that North Sea fish of all types were on borrowed time.

Mark Butcher recalls his time as a twelve-year-old during the school holidays, in the latter days of the market. He worked in 14a, the old wooden and metal Waveney Herring and Mackerel Market,

> 'When I worked there at 14a during the school holidays it was great. I was very puny and even to lift a box of fish was difficult. It really was a busy place. All those tatty old shacks that were falling to pieces. They were very, very old. 14a was very cold, very draughty. The floor was always damp and there were [fish] scales everywhere. I recall doing mainly plaice. I think it went up to London. I didn't do the winter!'

Waveney Herring and Mackerel Market opened in 1883. One end of the Herring Market had been destroyed during the Second World War and had been replaced by brick offices. The remaining part contained the Victorian-built offices which survived into the 1980s. By the time the present market was built it was too late. Fishing at Lowestoft was also in decline.

Some had already seen the writing on the wall. The Colne Shipping Company was one of the first to move away from fishing, although for its charismatic owner, Gordon Claridge, it could not have been an easy decision, as shipwright Albert Stone remembers,

> 'Mister Claridge was a very shrewd man, because on the onset of the standby vessels for the oil rigs he must have bought upwards of thirty to forty ships and we modified them from the very beginning. Very limited modification – but after the Piper Alpha disaster they became more and more sufficient [sic]. Some of the boats he bought [came] from Grimsby and Fleetwood. The Fleetwood ones were the biggest ones.'

From the 1970s, the EEC 'common pond' policy also meant that more vessels were fishing for an ever-decreasing stock. Despite later governmental dictates and restrictions, fish stocks continued to dwindle. Although eventually there was some improvement, EU quotas effectively throttled not just the North Sea fisherman but even the longshoreman's livelihood. The situation was not helped with the introduction of continental boats vacuuming everything off the sea floor.

In spite of the incessant red tape and increasing fuel prices, Gordon Claridge must have had some hopes of a return to North Sea fishing. Based in the last port in East Anglia where any substantial fishing was still taking place, Colne Shipping became the last major company to operate trawlers out of Lowestoft. Albert Stone remarks,

Early beam trawler (LT 266) on the left at Lowestoft, c.1980. (Author)

> 'In the end, all the fishing vessels became beam trawlers. In the onset of the 1970s and 1980s they were all what you would call 'side-winders', which were trawlers which would tow trawl doors. They were phased out when the Dutch became more proficient. Mr Claridge's heart was in the fish. He was purely and simply a fisherman.'

The Dutch favoured the beam trawler. The occasional Lowestoft beamer had already made an appearance and had started to fish out of the port in the late 1970s. As the 1980s wore on, they eventually replaced the old side-winders. The *St Anthony*, built in 1999 was the last Claridge boat dedicated solely to fishing – it was also the last boat built for Lowestoft and the last to leave the port.

LT 63 Willem Adriana, *Lowestoft Trawl Basin, c.1980. (Ernest Graystone)*

By 1980, Rossfish with its fish processing, freezing plant, and wholesale sales office next door on the Trawl Dock, was undoubtedly the largest employer on the Fish Market at Lowestoft. Ross was the first to fall. It had been in difficulties for some time, and although the factory itself remained relatively buoyant, its wholesale operation had declined. We didn't know it at the time but this would be the first of many changes affecting not only Lowestoft, but the whole of the east coast. The factory had been the brainchild of Gerry Raines, who was its first general manager following the takeover of Explorator in 1965. Now a director at Ross Grimsby, he retained a personal interest in Lowestoft and came down to see what could be rescued.

Every day when I was taking back auction prices and noting the drop in quality of the fish coming into the factory, it was clear that one way or another, sooner or later, we'd be out of a job. One day – I think it was the end of September 1981 – we were all called into Ken Coleman's office. Ken had taken over from Gerry Raines as general manager in 1968. We were then told that the Ross sales office at Lowestoft would be closing at the end of October. We were in effect given four weeks' notice. It had been a choice between Hull and Lowestoft and Lowestoft lost the toss. The factory itself followed with the remaining one hundred men and women being made redundant in March 1982. A little over ten years later, the factory, the office, and even the Trawl Market itself would go. An era was about to end.

For a while, the factory became the temporary home of those fish merchants who moved in while the belated new Fish Market was under construction. My old office, being one of the smaller in the building, still stood empty, but the sales office next door housed a massive walk-

Rossfish factory and office, Trawl Market, Lowestoft, November 1972. (Ernest Graystone)

in freezer. Some of my former colleagues set up fish businesses, something I had planned to do. Before long, however, my own plans were side-stepped and I started on a vastly different career well way from my fishing roots.

Over the last two decades of the 20th century, the decline of prime and whitefish stocks went hand-in-hand with general changes in eating habits. The high street fishmonger has been superseded by the supermarket. No longer can you find a Macfisheries or David Grieg in the town centres of Great Yarmouth, Ipswich or King's Lynn. Unlike Great Yarmouth, Lowestoft had no national chain of fishmongers; however, you could find Thain's fishmongers and restaurant at the Bridge, Cladingbowl & Stevens in Suffolk Road, and the North End Fish Stores in the High Street, among others, all traditional fishmongers. Apart from only a few smokehouses and a fishmonger in the Britten Centre, today almost all genuine cured fish has been replaced by 'smoked' fish that tastes more of dye than the smokehouse. As for the fried fish shop, the 'chippie'; not only is there competition from other fast-food outlets, but most have expanded beyond the traditional range of the three-penny bag of chips and a six-penny fish (the fag ash was always thrown in for free). Fish now has to compete with pies, sausages, mushy peas, drinks, chicken, spring rolls, beef cutlets and more. The queues waiting eagerly for the chippie to open have also gone.

Vera Barclay, whose father had a traditional fish and chip shop in the mid 1940s, recalls when they opened their shop it was surrounded by bombed out houses. They had to start from scratch,

> 'The shop was brick rubble. I was the one sent ahead to clean it before my parents moved in. The fireplace [frying range] was the iron fireplace – really rough. Scrubbed the floors and helped to get the rust off the metal work … Mum was the business woman. She handled the money.'

Vera also served in the shop,

> 'When we started the fish shop, we'd been trained in Norwich by my auntie's friend who ran a fish shop there. We went down and worked in their fish shop. He taught us how to make the batter, do the fish, how to fry it, how to do the chips. My mum did the filleting. Dad cleaned the potatoes. Mum had a special metal tray with holes to put the fish in so that the water drained away so that it wasn't sodden wet when you put it in the batter. Dad used to light the fires.
>
> 'When we first opened, meat was rationed, so we had queues. They came from Yarmouth and Gorleston to get our fish and chips – we couldn't believe it! Before we would open the queues would be up round Raglan Street!'

Now, Lowestoft struggled on in vain. In April 1995, when the EEC plaice quota earmarked for the east coast was given to the Dutch, fourteen Lowestoft boats were decommissioned. No longer did we hear the bell calling all to auction. No longer did we hear 'Let go the ropes' or 'Let go aft, let go for'rard.' By the end of the 1990s, it was obvious that even Lowestoft, the last great East Anglian fishing port, was in decline. Even if the fish did show any sign of returning in sufficient numbers, there wouldn't be enough boats left to catch them.

The EEC common fisheries policy, governmental red tape, ever-restricted quotas and the horrific rise in fuel costs meant that in the early years of the new millennium even Colne gave up any semblance of a fishing fleet at Lowestoft. Hamilton Dock, opened as an overspill from the Herring and Mackerel Market in 1906, became a mooring for longshore boats. When these declined, the dock became a marina.

The last fulltime fisherman working out of Yarmouth, Jason Clarke, announced in December 2009 that he was giving up fishing due to the frustrations of EU quotas. Thirty years ago, in the 1980s, Yarmouth had approximately twenty longshore boats. Only weeks after Mr Clarke's announcement, in January 2010, King's Lynn fishermen boycotted a governmental meeting held to discuss fish conservation plans. Only forty fishermen attended a similar meeting held days later in Lowestoft.

King's Lynn longshore boats still search for shellfish, travelling beyond the Wash as far as the Suffolk coast. In the months immediately following the First World War, King's Lynn had some twenty-five boats over fifteen tons unladen. Today, only small boats

LN 275 Sparkling Star *of King's Lynn, 1980s. (Ernest Graystone)*

LT 1005 St Anthony, *the last vessel built for Lowestoft, Trawl Basin, Lowestoft, May 2000. (Author)*

fish out of the port. Although the North End, that once great community of fishermen, went between the 1930s and the 1960s, several old fishing families still live in the area. Other than True's Yard itself, only a handful of buildings remain to give that authentic feel of King's Lynn fisherfolk and their heritage.

Lowestoft, that final bastion of the North Sea fishery in East Anglia, still survives – but only just. The Fish Market built in the 1980s is now home to a handful of wholesale fish merchants and is a shadow of its former self.

Will the fishing return? Will the herring make a comeback? I leave the last few words to Ernie Dewhurst, who has fished off the East Anglian coast since 1975,

> 'The fish [stocks] today has never been so good. The reason being … is that they catch the month's quota in the first week of the month. Fair enough, it's not like that all the time. I can remember back thirty years; we've gone out there and shot sixty to seventy long lines for a stone of fish.'

As for the herring,

> 'We used to do a lot of drifting, but that was just a waste of time. Hard work and no money. We used to catch lots, but when Sutton's folded in Yarmouth there was no processing. [When] the government put a ban on herring … all the processing plants closed down. It's like now, you can go out there and catch as much herring as you want but there's nowhere where you can sell it. Fifteen to twenty years ago you had a lot of merchants, now you have four and five left … The boats won't come back not like they were. You can't get the workforce, and the price doesn't warrant the effort put in and the expense – the price of diesel, the price of paint [and] gear.'

Further Reading

Bacon, J & S, *Suffolk Shoreline,* (SP, 1984)
Butcher, David, *The Driftermen*, (Tops'l Books, 1979)
Butcher, David, *Following the Fishing*, (Tops'l Books, 1979)
Butcher, David, *Living From The Sea*, (Tops'l Books, 1979)
Childs, Ernie, *My Little Book of Herring*, (Great Yarmouth Pottery)
Elliott, Colin, *Steam Fishermen*, (Tops'l Books, 1979)
Hedges, A.A.C., *Yarmouth is an Ancient Town*, (Great Yarmouth Borough Corporation 1959 and 1962)
Jones, Alan, *Life at 174 St Peter's Street – A Lowestoft Childhood*, (Lowestoft Heritage Workshop Centre)
King's Lynn official guides (various dates)
Malster, Robert, *Ipswich*, (Dalton, Lavenham, 1979)
Malster, Robert, *Lowestoft – east coast port* (Dalton, Lavenham, 1982)
Midgley, Patricia W., *The Northenders – a disappeared community*, (King's Lynn, 1987, 2nd ed.)
Mottram, R.H., *East Anglia – England's Eastern Province*, (Chapman & Hall, 1933)
Olsen almanacs
Prunier, Simone, *La Maison: The History of Prunier's* (Longman's, Green & Co., 1957)
Robb, Ian G., *Lowestoft – A History and Celebration* (Ottakar's/Frith, 2005)
Robb, Ian G., *Images of Lowestoft – the photographs of Christopher Wilson* (Sutton Publishing, 2002)
Universal British Directory – volumes 2 to 5 (circa 1792 to 1794)

Index

Also by Countryside Books